151
Advances in Polymer Science

Springer-Verlag Berlin Heidelberg GmbH

New Developments in Polymer Analytics II

Volume Editor: M. Schmidt

With contributions by
W. Köhler, R. Schäfer, S. S. Sheiko

Springer

This series presents critical reviews of the present and future trends in polymer and biopolymer science including chemistry, physical chemistry, physics and materials science. It is addressed to all scientists at universities and in industry who wish to keep abreast of advances in the topics covered.

As a rule, contributions are specially commissioned. The editors and publishers will, however, always be pleased to receive suggestions and supplementary information. Papers are accepted for „Advances in Polymer Science" in English.

In references Advances in Polymer Science is abbreviated Adv. Polym. Sci. and is cited as a journal.

Springer APS home page: http://link.springer.de/series/aps/ or
http://link.springer-ny.com/series/aps
Springer-Verlag home page: http://www.springer.de

ISSN 0065-3195
ISBN 978-3-662-15645-2 ISBN 978-3-540-48763-0 (eBook)
DOI 10.1007/978-3-540-48763-0

Library of Congress Catalog Card Number 61642

Originally published by Springer-Verlag in 2000
Softcover reprint of the hardcover 1st edition 2000

Typesetting: Data conversion by MEDIO, Berlin
Cover: E. Kirchner, Heidelberg
SPIN: 10706200 02/3020 - 5 4 3 2 1 0 - Printed on acid-free paper

Volume Editor

Prof. Manfred Schmidt
Institute of Physical Chemistry
University of Mainz
D- 55099 Mainz, FRG
E-mail: mschmidt@mzdmza.zdv.uni.mainz.de

Editorial Board

Preface

The two volumes "New Developments in Polymer Analytics" deal with recent progress in the characterization of polymers, mostly in solution but also at surfaces. Despite the fact that almost all of the described techniques are getting on in years, the contributions are expected to meet the readers interest because either the methods are newly applied to polymers or the instrumentation has achieved a major breakthrough leading to an enhanced utilizaton by polymer scientists.

The first volume concentrates on separation techniques. H. Pasch summarizes the recent successes of multi-dimensional chromatography in the characterization of copolymers. Both, chain length distribution and the compositional heterogeneity of copolymers are accessible. Capillary electrophoresis is widely and successfully utilized for the characterization of biopolymers, particular of DNA. It is only recently that the technique has been applied to the characterization of water soluble synthetic macromolecules. This contributrion of Grosche and Engelhardt focuses on the analysis of polyelectrolytes by capillary electophoresis. The last contribution of the first volume by Coelfen and Antonietti summarizes the achievements and pitfalls of field flow fractionation techniques. The major drawbacks in the instrumentation have been overcome in recent years and the "triple F techniques" are currently advancing to a powerful competitor to size exclusion chromatography.

The second volume starts with the introduction of a fascinating new technique developed by Köhler and Schäfer to monitor different averages of the Brownian diffusion coefficient of polymers in solution exhibiting a broad molar mass distribution. The technique is based on a scattering experiment by thermal grating, thus simultaneously monitoring Brownian and thermal diffusion of macromolecules. The last contribution by Sheiko addresses recent advances in atomic force microscopy. Besides some theoretical and experimental background the visualization of single macromolecules is discussed and complemented by the characterization and the patterning of surfaces.

Mainz, April 1999 Manfred Schmidt

Contents

Contents of Volume 150

New Developments in Polymer Analytics I

Volume Editor: M. Schmidt

Polymer Analysis by Thermal-Diffusion Forced Rayleigh Scattering

W. Köhler [1], R. Schäfer [2,3]

[1] Physikalisches Institut, Universität Bayreuth, D-95440 Bayreuth, Germany
e-mail: werner.koehler@uni-bayreuth.de
[2] Max-Planck-Institut für Polymerforschung, Ackermannweg 10, D-55021 Mainz, Germany
[3] Present address: IBM, Hechtsheimer Strasse 2, D-55131 Mainz, Germany
e-mail: rolfscha@de.ibm.com

The holographic grating technique of thermal-diffusion forced Rayleigh scattering (TDFRS) is used for the study of Fickian and thermal diffusion in simple liquids and polymer solutions. All three diffusion coefficients D_{th}, D, D_T and the Soret coefficient S_T can be obtained from a single experiment. Due to the short diffusion length of the order of a few μm, the whole system is very stable against perturbations like convection. TDFRS and photon correlation spectroscopy (PCS) are compared in detail. In case of polydisperse solutes, TDFRS does not suffer from the high scattering power of heavy components, which dominate the PCS correlation function. Quantities of interest for polymer analysis are distribution functions and averages of diffusion coefficients and molar masses. The weight distribution of the hydrodynamic radius follows directly from the heterodyne diffraction efficiency, without the need to resort to scaling relations. Pseudostochastic binary sequences with a broad power spectrum allow for the direct measurement of the linear response function, to which the individual molar masses contribute strictly concentration proportional, with a high spectral power density. Certain diffusive modes can be suppressed or enhanced by random binary sequences with colored power spectra, which are tailored to the specific experimental problem.

Keywords. Diffuson, thermal diffusion, forced Rayleigh scattering, polymer solutions, polydispersity, pseudostochastic noise

List of Abbreviations

a	normalized saturation amplitude of concentration grating
b	scaling exponent
C	color function in frequency domain
c	concentration (weight fraction)
$c'(\Gamma)$	concentration distribution as function of Γ
$c''(R_h)$	concentration distribution as function of R_h
c_0	equilibrium concentration
c_p	specific heat at constant pressure
c_q	concentration modulation depth
$c_{0,k}$	equilibrium concentration of k-th species
D	mutual mass diffusion coefficient

D_T	thermal diffusion coefficient
D_{th}	thermal diffusivity
d	fringe spacing of the grating
E	Fourier-transformed noise
E_s, E_c	Electric field amplitudes of signal and coherent background
$e(t)$	time dependent noise
$g(t)$	linear response function
$g_1(t)$	electric field autocorrelation function
$g_2(t)$	intensity autocorrelation function
$g_T(t)$	linear response function of temperature grating
H	Fourier-transformed total sample response function
$h(t)$	total sample response function
I	laser intensity (measured intensity for PCS)
$I(M)$	integral molar mass distribution
J	intensity measured by PMT\
L_C, L_T, L_G	universal constants for SNR otpimization
M	molar mass
M_n	number average molar mass
M_w	weight average molar mass
n	index of refraction
n_q	index of refraction modulation depth
$P(\Gamma)$	rate distribution function
P_r	power of readout laser
p_k	amplitude of *k*-th mode for multiexponential decay
q	grating or scattering vector
R_g	radius of gyration
R_h	hydrodynamic radius
S	source term in heat equation
S_{het}	heterodyne signal
S_{hom}	homodyne signal
S_T	Soret coefficient
s	sample thickness
T	temperature
T_q	temperature modulation depth
t	time
U	noise amplification or voltage
X	Fourier-transformed excitation
$x(t)$	general excitation function
$x[n]$	time-discrete binary excitation sequence
Y	Fourier-transformed signal
$y(t)$	general signal function
$y[n]$	time-discrete digitized signal
α	optical absorption coefficient
α	exponent for M in amplitude factor for multiexponential decay
η_0	solvent viscosity

ϕ	volume fraction
Γ	mass diffusion rate constant
κ	thermal conductivity
λ	wavelength of light
λ_r	wavelength for reading
λ_w	wavelength for writing
ρ	density
Σ	signal-to-noise ratio
τ	mass diffusion time constant
τ_k	mass diffusion time constant of k-th species
τ_p	excitation pulse length
τ_{th}	heat diffusion time constant
θ	scattering angle
ζ_{het}	heterodyne diffraction efficiency
ζ_{hom}	homodyne diffraction efficiency

1 Introduction

Thermal diffusion, also known as the Ludwig-Soret effect [1, 2], is the occurrence of mass transport driven by a temperature gradient in a multicomponent system. While the effect has been known since the last century, the investigation of the Ludwig-Soret effect in polymeric systems dates back to only the middle of this century, where Debye and Bueche employed a Clusius-Dickel thermogravitational column for polymer fractionation [3]. Langhammer [4] and recently Ecenarro [5, 6] utilized the same experimental technique, in which separation results from the interplay between thermal diffusion and convection. This results in a rather complicated experimental situation, which has been analyzed in detail by Tyrrell [7].

Diffusion cells which are heated from above and cooled from below represent another traditional experimental technique, which has been employed by e.g. Emery [8], Hoffman [9], Whitmore [10], Meyerhof and Nachtigall [11,12], and Giglio [13,14]. While in the earlier work the concentration gradient was analyzed layer by layer, optical beam deflection is now state-of-the-art, and good descriptions of such cells have been given by Kolodner [15] and Zhang [16]. Despite their conceptual simplicity, these experiments are not easy to perform, since convection must be avoided completely during the slow build-up of the concentration gradient.

Most data about the Ludwig-Soret effect of polymers in solution have been obtained from thermal field-flow fractionation (TFFF), developed by Giddings and coworkers [17,18]. TFFF is one member of the family of field-flow fractionation techniques, which are all characterized by a laminar flow of the polymer solution or colloidal suspension within a relatively narrow channel. An external field, which may be gravitation, cross-flow, or temperature as in TFFF, is applied perpendicular to the flow channel and biases the concentration distribution. Fractionation in TFFF results from the competition between thermal and Fickian diffusion, which is molar-mass-dependent, and from the varying velocity of the solution at different distances from the channel wall.

TFFF has been employed for the investigation of linear polymers in various solvents [19, 20, 21], copolymers [22, 23], branched polymers [23], and cross-linked microgels and block-copolymer micelles [24]. Kirkland coupled TFFF with an online viscosity detector [25].

So far, there is a considerable uncertainty about the Soret coefficients in the literature, and there is only one system, namely polystyrene (PS) in toluene, which has been investigated by a substantial number of different authors with different experimental techniques, and for which a reasonable agreement has been achieved [6, 11, 12, 19, 21].

In 1978, Thyagarajan and Lallemand observed a fast heat and a slow concentration mode in forced Rayleigh scattering experiments on a mixture of CS_2 and ethanol [27]. The slow mode stems from a concentration grating, which is induced by the holographic temperature grating that is written into the solution. During the last years, this optical technique, which one might call thermal-diffusion forced Rayleigh scattering (TDFRS), has been employed in our laboratory for the study thermal diffusion and mass diffusion of polymers in solution [28].

The concept of TDFRS is very simple. A holographic interference grating is written into the solution. It heats the sample with a spatially periodic pattern and creates a temperature grating. Optical absorption of the usually transparent solvents is achieved by the addition of some inert dye. The temperature gradients within the temperature grating give rise to thermal diffusion, and a secondary concentration grating builds up. Both the temperature and the concentration grating are accompanied by a refractive index grating, which can be read by Bragg diffraction of a readout laser beam.

TDFRS allows for experiments on a micro- to mesoscopic length scale with short subsecond diffusion time constants, which eliminate almost all convection problems. There is no permanent bleaching of the dye as in related forced Rayleigh scattering experiments with photochromic markers [29, 30] and no chemical modification of the polymer. Furthermore, the perturbations are extremely weak, and the solution stays close to thermal equilibrium.

Besides the investigation of thermal diffusion, TDFRS is an interesting tool for the study of mass diffusion processes. The Ludwig-Soret effect merely aids in the preparation of an initial nonequilibrium state with a single well-defined q-vector, whose time evolution is later observed. By employing different time patterns for the excitation, it is possible to tune the signal contribution of different species, which may be of different chemical composition or different molar mass. This is not possible in case of photon correlation spectroscopy (PCS), where the correlation functions are always dominated by the strongly scattering heavy components. It turns out that the concentration part of the linear response function of the sample is of particular interest, since the statistical weights of the individual contributions are completely independent of their respective molar mass. Since the advent of fast digital correlation and Fourier transform techniques, the direct measurement of the response function of a linear system with broad-band excitations of high spectral power density has been feasible.

In this contribution, the experimental concept and a phenomenological description of signal generation in TDFRS will first be developed. Then, some experiments on simple liquids will be discussed. After the extension of the model to polydisperse solutes, TDFRS will be applied to polymer analysis, where the quantities of interest are diffusion coefficients, molar mass distributions and molar mass averages. In the last chapter of this article, it will be shown how pseudostochastic noise-like excitation patterns can be employed in TDFRS for the direct measurement of the linear response function and for the selective excitation of certain frequency ranges of interest by means of tailored pseudostochastic binary sequences.

2 Experimental Setup and Principles of Measurement

Before a phenomenological model for signal generation in TDFRS will be developed in the next section, some experimental aspects and the principles of signal detection and readout of the holographic phase grating need to be outlined. Most of the arguments are very general and apply to a wider class of holographic grating experiments.

For the moment, it is sufficient to know that the sample response, which is the time dependent diffraction efficiency after switching off the optical grating, contains at least a fast contribution from heat and a slow one from mass diffusion. The corresponding diffusion time constants depend on the grating constant and are typically of the order of 10 μs and 100 ms, respectively.

2.1 Setup

The main parts of the TDFRS setup and some experimental aspects will be discussed in this section without going into details of data evaluation and signal modelling. A derivation of the equations for a phenomenological description will be deferred until later.

A TDFRS setup is, in its main components, almost identical to standard forced Rayleigh scattering (FRS) setups as described in many publications [28, 31, 32, 33, 34]. The symmetric arrangement as employed for the more recent work in our group is sketched in Fig. 1.

The whole setup is mounted on a floating optical table. An argon-ion laser operating at 488 nm serves for writing. Its beam is spatially filtered and expanded to a diameter of 5-10 mm. The polarization is perpendicular to the optical table. The beam is split into two halves of equal intensity and approximately symmetric optical paths. After the beam splitter, there are mirrors mounted on piezo ce-

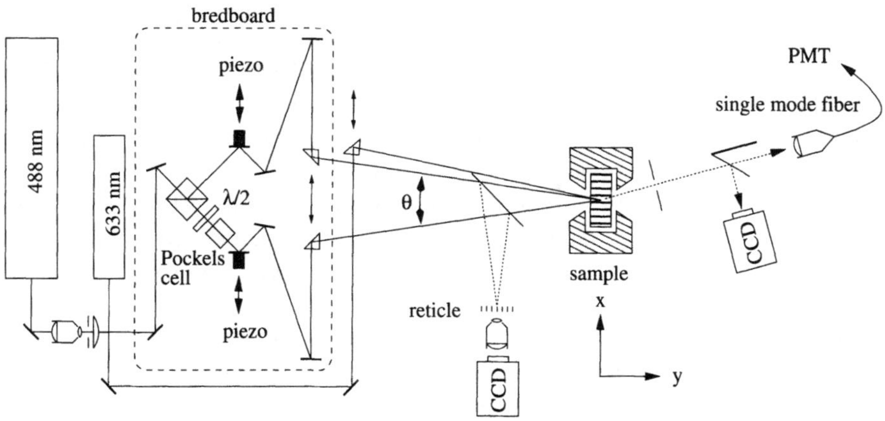

Fig. 1. Sketch of the experimental setup as employed in the more recent work. All critical parts are mounted on a separate breadboard for increased mechanical stiffness

ramics, which are used for phase stabilization and phase modulation of the grating. After another two adjustable mirrors, whose purpose is merely to simplify alignment after changes of the angle of intersection θ, the beams are steered towards the sample by two adjustable prisms. Once the mirrors are aligned, they need not be touched anymore, and θ is changed by adjusting the prisms as indicated by the arrows. A half wave plate and a Pockels cell are inserted into one of the beams right after the beam splitter. They serve for 180°-phase modulation, which is the optimum switching technique for heterodyne experiments. For maximum mechanical stiffness, all optical components for the split writing beams are mounted on a separate breadboard with rails and carriers, but without posts, since the vibration amplitude of these components must be kept below a few nanometers.

The angle of intersection θ is typically of the order of a few degrees or below, which is difficult to measure to the required precision of better than 0.5 %. Instead, as shown in Fig. 1, the fringe spacing of the grating is measured directly by projecting the grating together with a calibrated microscope reticle with a 25 μm scale onto a CCD camera, whose signal is displayed on a monitor and simultaneously fed into a video digitizer board [35].

The cuvette with the sample resides inside a temperature controlled brass housing and can be adjusted in both the x- and z-directions orthogonal to the optical axis. In most cases, optical flow cells with path lengths of 100 or 200 μm have been employed.

The diffraction efficiency of the grating is read by a HeNe-laser at 632.8 nm wavelength, which, because of the thick grating conditions, must be adjusted to its Bragg-angle. In our earlier work, a pinhole and bandpass filter arrangement was used to separate the diffracted beam from unwanted stray light and fluorescence [28, 34, 35, 36, 37]. Superior incoherent background suppression is achieved by the single-mode optical fiber shown in Fig. 1. Correctly aligned, the fiber allows only the diffracted beam and coherent stray light to propagate to the detector. The coupling efficiency of the fiber is better than 80%, and the full background coherence is ideal for heterodyne detection. The fiber is connected to a photo multiplier tube (PMT) operating in photon counting mode, which is hooked up to a counter board plugged into a personal computer. Count rates above $10^5\ s^{-1}$ require proper linearization [28].

The shortest integration time of the counter is currently set to 1 μs, corresponding to a maximum sampling rate of 1 MHz. The actual time resolution is limited by the high voltage driver of the Pockels cell with approximately 20 μs switching time.

Different switching techniques can be employed for the grating. As will be discussed later, it is advantageous to keep the spatial average of the light intensity on the sample constant in time. If both the heterodyne and the homodyne signal are needed (see below), the grating amplitude must be switched between zero, corresponding to even illumination, and some finite value. This is accomplished by a 90°-rotation of the polarization of one of the writing beams. If only the linear heterodyne signal is of interest, the amplitude of the grating is switched between −1 and +1, corresponding to a 180°-phase jump. Both the Pockels cell and the piezo can be used for this purpose, but, unfortunately, both require subtle corrections for small amplitudes or phase drifts.

2.2
Homodyne versus Heterodyne Detection Schemes

A distinction between homodyne and heterodyne detection must be made in optical scattering and diffraction experiments. Without careful treatment of the background, there is always the risk of mixed or unknown coherence conditions, and the diffusion coefficient determined from such data may be off by a factor of two. At least for the signal and background levels present in TDFRS, heterodyne detection is always superior to homodyne, especially since the heterodyne signal, contrary to the homodyne one, turns out to be very stable against perturbations and systematic errors. Even under nearly perfect homodyne conditions the tail of the decay curve is almost unavoidably heterodyne [34].

The intensity J as seen by the detector contains both homodyne and heterodyne contributions [34, 38, 39]:

$$J = \left| E_c + E_s e^{i\phi} \right|^2 + E_{inc}^2 = E_s^2 + 2E_s E_c \cos\phi + E_c^2 + E_{inc}^2 \quad . \tag{1}$$

E_s is the electric field amplitude of the diffracted beam. E_c and E_{inc} are the coherent and incoherent electric field amplitudes of the background intensity, respectively. ϕ is the phase shift between the signal and the coherent background, and the phase of E_c is arbitrarily chosen to be zero. For convenience, the proportionality factor between E^2 and J is set to unity. $S_{hom} = E_s^2$ is the homodyne and $S_{het} = 2\,E_c\,E_s \cos\phi$ the heterodyne signal. The total background is $J_b = E_c^2 + E_{inc}^2$.

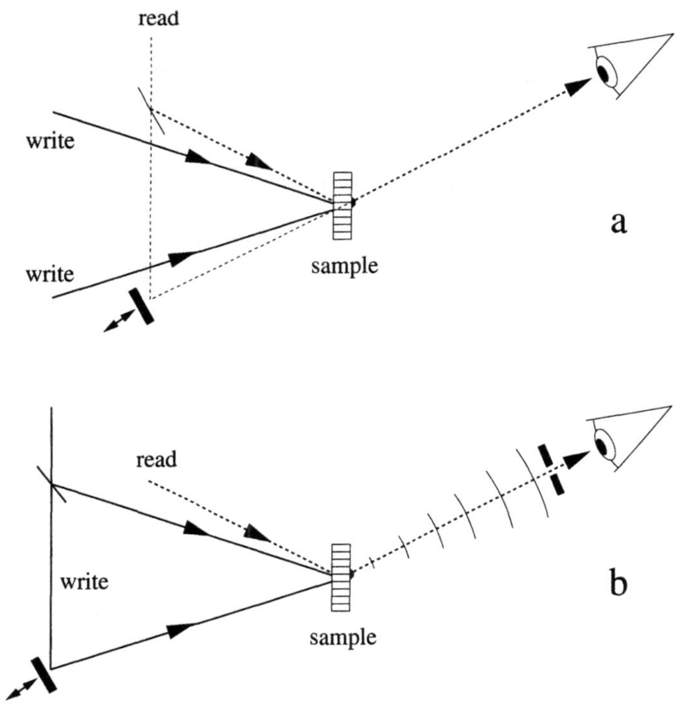

Fig. 2. Heterodyne detection schemes: phase control of the reference beam (**a**) and of the diffracted beam with the reference provided by local oscillator (**b**). From Ref. [34]

The homodyne and heterodyne signals can be separated if two measurements with a phase shift of π between background and signal are combined according to

$$S_{hom} = \frac{1}{2}\left(J_{\phi} + J_{\phi+\pi}\right) - J_b \ , \tag{2}$$

$$S_{het} = \frac{1}{2}\left(J_{\phi} - J_{\phi+\pi}\right) \ . \tag{3}$$

In principle, the coherent background can be supplied as shown in Fig. 2a, where the reference beam is directed along the diffracted beam. The phase of the reference is directly controlled by means of a piezo-mounted mirror.

An alternative scheme, which is much easier to handle, is shown in Fig. 2b, where dust or scratches on the cuvette windows serve as local oscillators for the generation of the reference wave. Since the phase of the reference is now fixed with respect to the sample, the phase of the diffracted beam must be controlled instead by shifting the phase of one of the writing beams, and hence of the grating. Due to the first order Bragg condition, the phase shift of the grating translates directly to the phase shift of the diffracted beam.

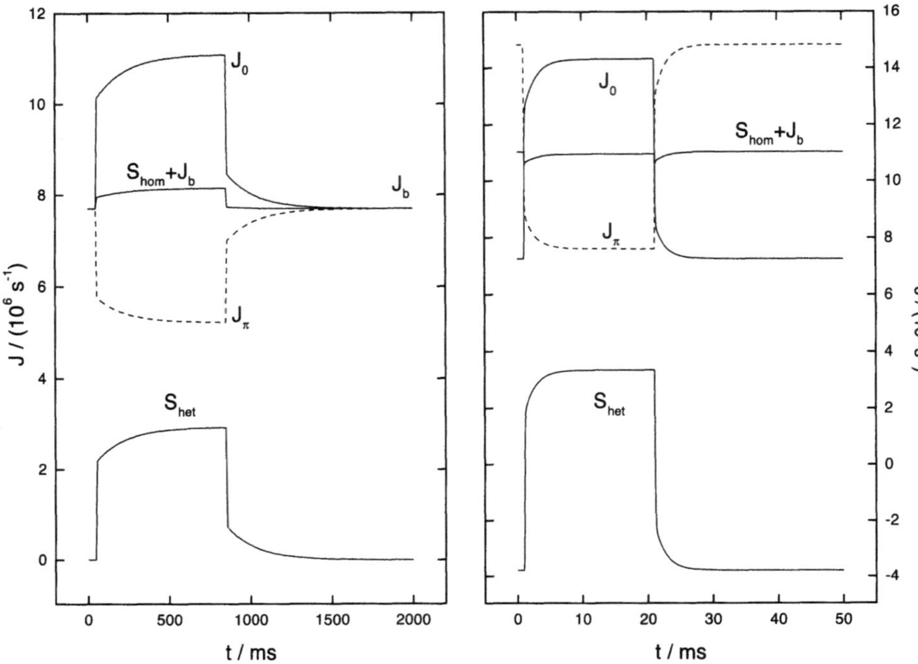

Fig. 3. Homodyne/heterodyne separation: measured intensities for step excitation and phase angles $\phi = 0$ and $\phi = \pi$, together with extracted homodyne and heterodyne signals. **Left**: 90° rotation of the polarization, sample PS/toluene. **Right**: 180° rotation of the polarization, sample toluene/n-hexane From Ref. [34]

Figure 3 shows the signals measured for $\phi = 0$ and $\phi = \pi$ as well as the separation into homodyne and heterodyne components for 90° and 180° rotation of the polarization. The superior quality of the heterodyne signal is obvious, and the homodyne signal is only usable in the case of 90° rotation.

2.3 Active Phase-Tracking

Ideally, there is no phase shift between the reference and the diffracted beam ($\phi = 0$), and, since TDFRS is completely nondestructive without dye bleaching, the signal can be accumulated over almost arbitrary times. In order to maximize the heterodyne signal amplitude, some means for phase adjustment and stabilization are needed. Without such active phase-tracking, ϕ would have some arbitrary value and would slowly drift away due to the almost unavoidable slow thermal drift of the whole setup.

An elegant phase-tracking and stabilization procedure has been developed, which utilizes the fast signal contribution from the heat mode with a relaxation time of the order of 10 μs, which is always present in TDFRS measurements [34].

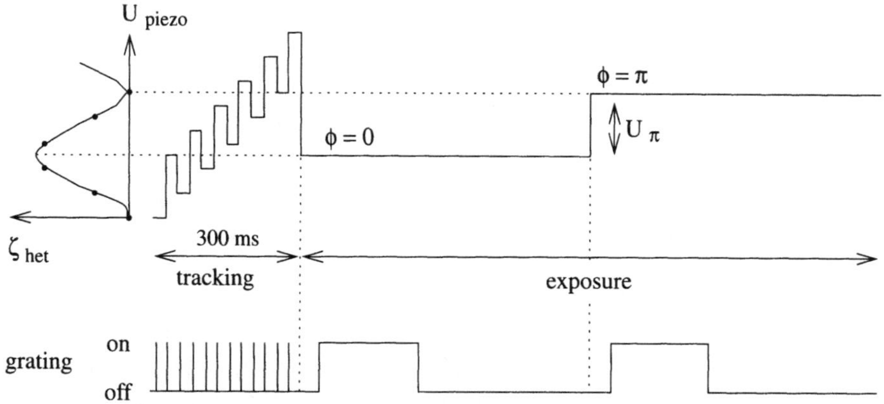

Fig. 4. Timing diagram for piezo mirror and Pockels cell with phase tracking and exposure. Details can be found in Ref. [34]. From Ref. [34]

For a given phase angle ϕ of the grating, corresponding to a certain position of the upper piezo mirror in Fig. 1, the heterodyne amplitude of the fast thermal signal is measured by integrating the diffracted intensity over a short exposure time of approximately 5 ms to obtain J_ϕ, then shifting the phase of the grating by π and exposing again to measure $J_{\phi+\pi}$. S_{het} is obtained according to Eq. 3. The piezo mirror is ramped such that ϕ varies by 2π, and S_{het} is measured at 5 to 10 intermediate positions as described. When plotted as a function of the piezo position, one period of a sinusoidal is obtained, from which the new ($\phi = 0$)-position, which corresponds to the largest heterodyne amplitude, can be calculated by means of a nonlinear least-squares fit [40]. The whole procedure requires only a few hundred milliseconds and is repeated approximately every minute to ensure a stable and well defined phase for optimum heterodyne signals over arbitrary integration times. In between, repeated exposures of the sample with $\phi = 0$ and $\phi = \pi$ are accumulated, the number depending on the mass diffusion time constant. The whole cycle can be repeated as long as needed for an aimed-at signal-to-noise ratio. The corresponding timing diagram is shown in Fig.4. The whole experiment, including timing, data acquisition and processing, phase stabilization, temperature control, and grating measurement is controlled by a single personal computer under the Linux operating system.

2.4 Optimization

Some constraints apply for the measurement of mass and thermal diffusion by TDFRS, which originate from excessive sample heating at high laser powers, the resulting onset of convection, and the need to avoid boundary effects at the cuvette windows when the grating constant becomes comparable to the sample thickness. The problem of optimization of the experimental boundary conditions has been analyzed in detail in Ref. [34], and therefore only the main results will briefly be summarized.

It can be shown that the best achievable heterodyne signal-to-noise ratio in the limit of 'infinite' coherent background intensity, $\Sigma^{opt}_{het,\infty}$ is given by

$$\Sigma^{opt}_{het,\infty} = \lambda_r^{-1}\left[8\beta\frac{P_r}{h\nu_r}\phi t_c\right]^{1/2}\left[\frac{\sqrt{\pi}}{2}L_c L_G^2 L_T^{1/2}\right]^{2/3}\left[|S_T c(1-c)(\frac{\partial n}{\partial c})_{p,T}|\frac{(\eta D)^{2/3}}{\kappa_s}\right] \quad (4)$$

In the above equation, $\Sigma^{opt}_{het,\infty}$ is written as a product of three terms. The first one contains only the parameters of the experimental setup, which are the variables accessible to the experimentalist for SNR improvement: λ_r is the readout wavelength and ν_r the corresponding frequency, $\beta = E_c^2/(E_c^2+E_{inc}^2)$ the degree of background coherence, P_r the power of the readout laser, ϕ the quantum yield of the detection system, and t_c the integrated sampling time per time channel.

The second term contains the constraint parameters, which may be regarded (within certain limits) as universal constants. They have been estimated in Ref. [34] to be:

$$\begin{aligned} L_C &\approx 200\ \mathrm{m}^2 s^{-1} \\ L_T &\approx 200\ \mathrm{W}^2\mathrm{m}^2 \\ \mathrm{L_G} &\approx 0.1\ . \end{aligned} \quad (5)$$

All sample specific quantities are found within the last term. η is the solution viscosity, D the diffusion coefficient, κ_s the thermal conductivity, S_T the Soret coefficient, and $(\partial n / \partial c)_{T,p}$ the concentration derivative of the refractive index at constant temperature and pressure.

In order to achieve $\Sigma^{opt}_{het,\infty}$ the measurements require both an optimized sample thickness s_{opt} and fringe spacing d_{opt}:

$$s_{opt} = \left(4\pi^2\eta D\frac{L_e}{L_T L_G}\right)^{1/3} \quad (6)$$

$$d_{opt} = s_{opt}L_G\ . \quad (7)$$

For the measurement of polystyrene with a molar mass $M_w = 410$ kg/mol in toluene, $s_{opt} \approx 230\ \mu$m and $d_{opt} \approx 23\ \mu$m is obtained from Eqs. 5–7 [34].

Of special interest are dilute polymer solutions, where the thermal diffusion coefficient, $D_T = S_T D$, is independent of molar mass [11, 21]. For the diffusion coefficient D, a scaling law $D \propto M^{-a}$ holds, with the exponent for linear flexible polymers ranging from $a = 0.5$ for θ-conditions to $a \approx 0.6$ for good solvents. Thus, the optimum SNR for a given polymer is almost independent of molar mass: $\Sigma^{opt}_{het,\infty} \approx M^{1/6 \ldots 1/5}$. This is an important feature, as it makes the entire molar mass range, which may cover many decades, accessible with approximately constant sensitivity- in sharp contrast to other optical scattering techniques.

2.5 Sample Preparation

Sample preparation is done in a similar way as for light scattering experiments. The main difference is that usually some inert dye must be added for absorption and thermalization of the energy from the light field. Typically, the optical density at the writing wavelength is adjusted to approximately 2 for a path lenght of 1 cm. In the case of polymer solutions, the dye is first added to the solvent, then the polymer is dissolved. After filtering several times through a 0.2 μm filter (Millex FGS, Millipore), the solution or mixture is filtered directly into the optical flow cell with a path length of 100 or 200 μm (Hellma). The volume needed for the experiment is below 2 μl.

The purpose of the dye is to provide optical absorption at the writing wavelength, but ideally not at the readout wavelength, of the otherwise transparent liquids. While some absorption at λ_r is not a principal problem, it results in avoidable sample heating.

There should be no chemistry involved between the dye and the polymer, and the dye should not undergo photo-reactions. For organic solvents, quinizarin fulfils both these requirements, and its absorption spectrum is shown Fig. 5.

Whether the dye is inert or whether it generates some time dependent signal by itself (e.g. through a photoreaction or transition into a triplet state) is easily checked from a measurement of the dyed solvent without polymer solute.

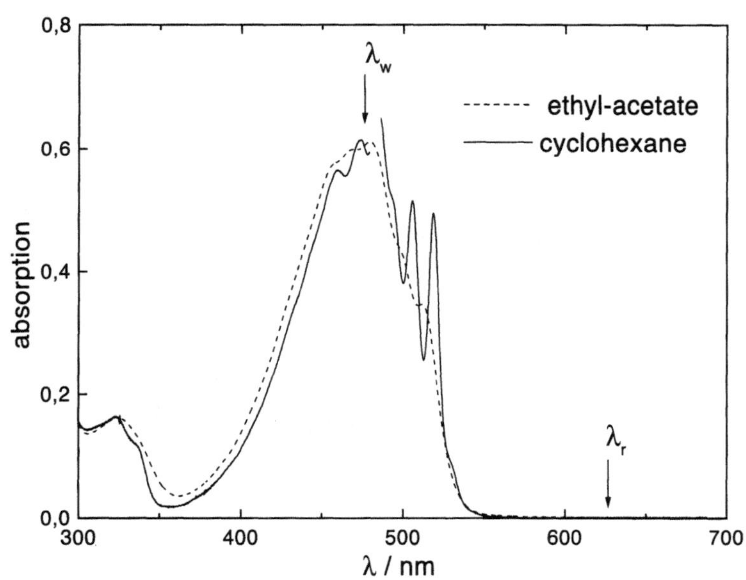

Fig. 5. Absorption spectrum of quinizarin in ethyl-acetate and cyclohexane. The dye is ideally matched to writing at 488 nm and readout at 633 nm

It should be noted that for aqueous systems a suitable dye was not easily found, since many water-soluble dyes apparently undergo a photoreaction in the presence of triplet oxygen, which is photochemically transformed into its reactive singlet form due to sensibilization by the dye [41]. After degassing with argon, good results for aqueous systems have been achieved with Orange I [42].

2.6 Contrast Factors

The contrast factors $(\partial n / \partial T)_{c,p}$ and $(\partial n / \partial c)_{T,p}$, taken at the wavelength of the readout laser, are the only quantities which must be measured in separate experiments. For this purpose, a Michelson interferometer has been developed, whose mirrors can be scanned over $\lambda / 2$ [43] (Fig. 6).

During the scan, the interference signal as detected by the photodiode goes through one period of a cosine, whose phase can be obtained from a nonlinear least-squares fit of

$$I_{diode} = A\left(1+\cos\left(2\pi\frac{U}{U_{2\pi}} - \Phi\right)\right) + B \ . \tag{8}$$

A and B are constants, and $U_{2\pi}$ is the voltage for a phase shift of 2π. A change of Φ is directly proportional to the change in the refractive index of the sample.

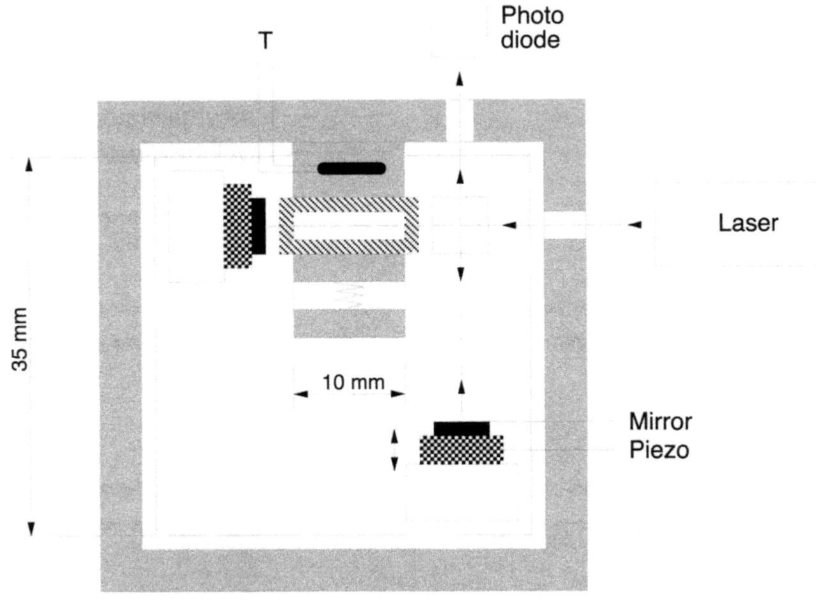

Fig. 6. Scanning Michelson interferometer as employed for the measurement of $(\partial n/\partial T)_{c,p}$. From Ref. [43]

Details of the procedure and necessary corrections for the thermal expansion of the cuvette and the ambient air can be found in [43].

Since the interferometer used for $(\partial n / \partial T)_{c,p}$ measurement is heated completely, and not just the cuvette, it has been made out of Zerodur (Schott, Mainz), which has a negligible thermal expansion coefficient. Precise values of the refractive index increments are crucial for the determination of the thermal diffusion coefficient and the Soret coefficient. The accuracy achieved for $(\partial n / \partial c)_{T,p}$ is usually better than 1 %, and the accuracy of $(\partial n / \partial T)_{c,p}$ better than 0.1 %.

3 Phenomenology of TDFRS

3.1 Writing the Temperature Grating

The phenomenological description of signal generation in TDFRS is, in principle, straightforward [27, 28, 33, 44, 45]. The primary source of excitation is the optical interference grating formed by the two writing beams, which intersect under an angel θ inside the sample:

$$I(x,t) = I_0 + I_q(t)e^{iqx} \tag{9}$$

The geometry of the setup is sketched in Fig. 7.

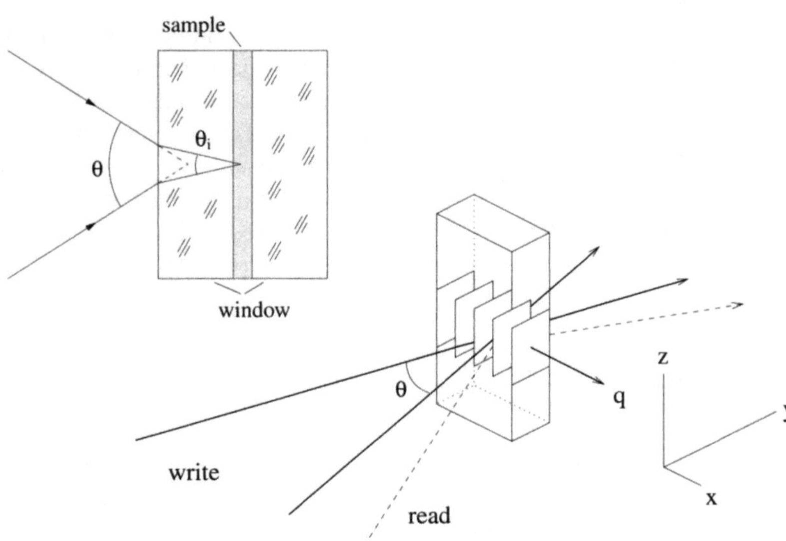

Fig. 7. Sketch of the geometry of the TDFRS-setup with the two writing beams intersecting under an external angle θ. Also shown is a cross section through the sample cell

The optical axis defines the y-direction. The x-axis points along the direction of the grating vector , whose absolute value is

$$q = \frac{4\pi}{\lambda} \sin\frac{\theta}{2} \ . \tag{10}$$

λ is, due to the plane-parallel geometry of the sample cell, the wavelength of light in air. With a plane-parallel geometry of the sample cell and the optical axis perpendicular to the windows, the refractive index n of the sample does not appear in Eq. (9), since, according to Snell's law,

$$n \sin\frac{\theta_i}{2} = n_a \sin\frac{\theta}{2} \ . \tag{11}$$

$n_a \approx 1$ is the refractive index of air and η_i the angle of intersection inside the sample as defined in Fig. 7. For a different cell geometry, Eq. (10) may need corrections for the refractive index of the sample. I_0 is the constant spatial intensity average, which is the incoherent sum of the intensities $I_0/2$ of the two writing beams. $I_q(t)$ is the amplitude or modulation depth of the grating, and we will discuss later how the time dependence of $I_q(t)$ can be tailored to meet specific experimental needs.

The complex notation in Eq. (9) indicates that the grating is characterized by both an amplitude and a phase factor, and $I_q(t)$ may also be a complex quantity containing both amplitude and phase modulation. For heterodyne detection schemes, both contributions are of importance. In most experiments, but not in all, 180°-phase switching, corresponding to $\pm I_0$-amplitude switching of the interference grating, has been employed.

Both finite-size effects of the beam diameter and exponential intensity decay due to absorption within the sample have been neglected in Eq. (9). A purely one-dimensional treatment of the problem, as implied in Fig. 7 and as usually employed in the literature, is not always adequate and can only be justified if the experimental boundary conditions are carefully chosen.

The direct coupling mechanism between the optical grating and the sample is the absorption of energy from the light field, which results in a periodic heating with the same q-vector as the optical grating. To describe this temperature grating, we start with the heat equation

$$\frac{\partial T(\vec{r},t)}{\partial t} = D_{th} \Delta T(\vec{r},t) + \frac{\dot{Q}(\vec{r},t)}{\rho c_p} \ . \tag{12}$$

$T(\vec{r},t)$ is the spatial and temporal temperature distribution, D_{th} the thermal diffusivity, ρ the density, c_p the specific heat at constant pressure, and $\dot{Q}(\vec{r},t)$ the local heat production per volume. A general solution of Eq. (12) with the appropriate boundary conditions, including thermal conductivity of the cell windows and heat transition to the ambient air, can be a challenging task. The whole problem is simplified, since the experiment is set up in such a way that it only probes Bragg diffraction from the refractive index modulations with the grating vector $\vec{q}$ as defined in Eq. (10) and Fig. 7.

There are, however, at least two other characteristic length scales besides the fringe spacing of the grating: the sample thickness and the writing beam diameter. While, within the assumption of a linear response with negligible temperature dependence of D_{th}, ρ, and c_p, these length scales do not directly interfere with the grating, they lead to the formation of a time-dependent thermal lens. This thermal lens is responsible for a slight time-dependent beam displacement and defocusing, which may show up in the recorded signal. Furthermore, background light scattered from dust or scratches on the cuvette entrance window has to travel through the thermal lens, where it experiences a difficult-to-handle time-dependent phase shift before it coherently adds to the diffracted signal.

Since the length scales associated with the thermal lens are on the order of 10 to 1000 times the grating constant, their characteristic time scale interferes with polymer diffusion within the grating. Such thermal lensing has been ignored in many FRS experiments with pulsed laser excitation [27,46] and requires a rather complicated treatment. A detailed discussion of transient heating and finite size effects for the measurement of thermal diffusivities of liquids can be found in Ref. [47].

All these problems can be avoided, if the writing laser beams are not switched on and off but the phase of the interference grating is modulated instead according to Eq. (9), while keeping the average intensity distribution within the laser spot constant. Under these conditions, it can be justified to employ Eq. (12) in its one-dimensional form [27,28,45]:

$$\frac{\partial T(x,t)}{\partial t} = D_{th} \frac{\partial^2}{x^2} T(x,t) + S(x,t) \ . \tag{13}$$

The source term $S(x,t)$ is given by the absorbed energy according to Eq. (9):

$$S(x,t) = \frac{\alpha}{\rho c_p} I(x,t) = S_0 + S_q(t) e^{iqx} \ . \tag{14}$$

For the sake of simplicity, a quantum yield of unity has been assumed for the thermalization of the absorbed light. Eq. (14) is solved by

$$T(x,t) = T_0 + T_m(t) + T_q(t) e^{iqx} \ . \tag{15}$$

T_0 is the initial sample temperature. The amplitude $T_q(t)$ of the temperature grating is expressed as a linear response for arbitrary excitations $S_q(t) = \alpha\,(\rho\, c_p)^{-1}\, I_q(t)$:

$$T_q(t) = \int_{-\infty}^{t} dt' S_q(t') e^{-(t-t')/\tau_{th}} \ . \tag{16}$$

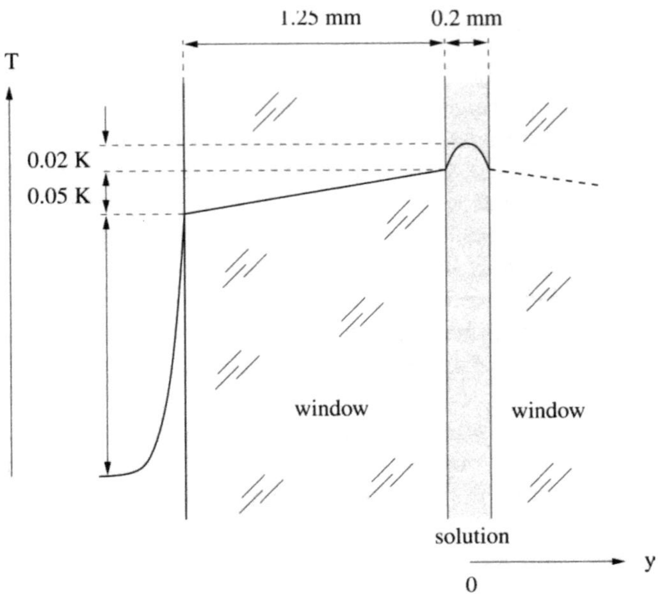

Fig. 8. Temperature distribution within the cuvette in case of one-dimensional heat transport along the optical axis for typical experimental conditions ($\alpha \approx 20^{-1}, I_0 \approx 25$ mW cm^{-2}). From Ref. [34]

$\tau_{th} = (D_{th}q^2)^{-1}$ is the heat diffusion time. An instantaneous thermalization of the optical excitation has been assumed.

Of course, the mean sample temperature

$$T_m(t) = \frac{\alpha}{\rho c_p} I_0 t \tag{17}$$

does not grow forever, as might be suggested by Eq. (17), but reaches instead some stationary value, as the sample permanently loses heat to its environment. T_m is responsible for the slight thermal lensing discussed above and, strictly speaking, not spatially constant.

The effect of sample heating has been treated in detail in Ref. [34], and the temperature rise within the sample and the cuvette is summarized for typical experimental conditions in Fig. 8. From interferometric measurements the overall temperature increase of the cuvette has been obtained to approximately 0.1 K, but the non-uniformity within the 200 μm thick sample is only 0.02 K, as shown in Fig. 8.

3.2 Formation of the Concentration and Refractive Index Gratings

Concentration grating: Due to the Ludwig-Soret effect, the temperature grating is the driving force for a secondary concentration grating, which starts to build up and is superimposed upon the thermal one. Its temporal and spatial evolution is obtained from the one-dimensional form of the extended diffusion equation

$$\frac{\partial c(x,t)}{\partial t} = D\frac{\partial^2}{x^2}c(x,t) + D_T c_0(1-c_0)\frac{\partial^2}{x^2}T(t,x) \, , \tag{18}$$

which is solved by

$$c(x,t) = c_0 + c_q(t)e^{iqx} \tag{19}$$

with

$$c_q(t) = -q^2 D_T c_0(1-c_0)\int_{-\infty}^{t} dt' T_q(t')e^{-(t-t')/\tau} \, . \tag{20}$$

$\tau = (Dq^2)^{-1}$ is the collective diffusion time constant, D_T the thermal diffusion coefficient. In Eq. (18), the low modulation depth approximation $c_q(t) \ll c_0$, resulting in $c(x,t)(1-c(x,t)) \approx c_0(1-c_0)$, has been made, which is valid for experiments not too close to phase transitions. Eqs. (16) and (20) provide the framework for the computation of the temperature and concentration grating following an arbitrary optical excitation.

Refractive index grating and heterodyne detection: For a sample transparent at the readout wavelength, a pure refractive index or phase grating builds up, with contributions from both the temperature and the concentration gratings. In analogy to Eqs. (15) and (19), it is expressed as

$$n(x,t) - n_0 = n_q(t)e^{iqx} = \left[\left(\frac{\partial n}{\partial T}\right)_{c,p} T_q(t) + \left(\frac{\partial n}{\partial c}\right)_{T,p} c_q(t)\right]e^{iqx} \tag{21}$$

n is the refractive index at the readout wavelength. The respective contrast factors are the temperature derivative of n at constant composition and pressure, $(\partial n/\partial T)_{c,p}$, and the concentration derivative of n at constant temperature and pressure, $(\partial n/\partial c)_{T,p}$. Pure phase gratings have been assumed for the present discussion, but absorptive contributions can be incorporated without problem [39].

Generally, a distinction is made between thin and thick gratings according to the criterion[39]

$$Q = \frac{2\pi s \lambda_r}{d^2 n} \, . \tag{22}$$

λ_r is the readout wavelength and s the sample thickness. $Q \ll 1$ indicates thin, $Q \gg 1$ thick grating conditions, which are prevalent in the present case with $Q \approx 5$. An extensive treatment of the diffraction theory of phase gratings has been given by Kogelnik [48]. All experiments discussed here have been conducted within the weak modulation depth limit, where the heterodyne or electric field diffraction efficiency $\zeta_{het}(t)$ is simply proportional to the refractive index modulation depth:

$$\zeta_{het}(t) \propto n_q(t) \; . \tag{23}$$

$\zeta_{het}(t)$ can be complex to account for phase shifts. The homodyne diffraction efficiency, which is measured in the absence of coherent background, is proportional to $|n_q(t)|^2$.

Now, all steps can be combined to calculate the heterodyne diffraction efficiency from Eqs. 9, 16, 20, 21, and 23. After normalization to the diffraction efficiency of the steady state amplitude of the temperature grating, one arrives at

$$\zeta_{het}(t) = \delta T^{-1} \int_{-\infty}^{t} dt' g(t-t') T_q(t') \; . \tag{24}$$

$\delta T = \tau_{th} S_0$, and all information about the sample is contained within the memory or linear response function

$$g(t) = \begin{cases} \delta(t) - (\frac{\partial n}{\partial c})_{p,T} (\frac{\partial n}{\partial T})^{-1}_{p,c} q^2 D_T c_0 (1-c_0) e^{-t/\tau} & t \geq 0 \\ 0 & t < 0 \end{cases} \; . \tag{25}$$

The δ-function accounts for the fast contribution from the temperature grating. The normalization to the amplitude of the temperature grating in Eq. (24) takes away the need for difficult absolute intensity measurements for the determination of D_T and S_T, which otherwise would be necessary. The only quantities that must be obtained from separate measurements are the two contrast factors $(\partial n/\partial T)_{c,p}$ and $(\partial n/\partial c)_{T,p}$. All the transport coefficients D_{th}, D, D_T, and S_T can be extracted from the sample response to suitably chosen excitation patterns.

Equation (25) holds for monodisperse samples with a single diffusion time constant τ, and the diffraction efficiency in response to the most fundamental excitation, a step function where the grating amplitude is switched from 0 to 1 at $t = 0$, is [27, 28, 35, 45]

$$\zeta_{het}(t) = 1 - e^{-t/\tau_{th}} - \left(\frac{\partial n}{\partial c}\right)_{p,T} \left(\frac{\partial n}{\partial T}\right)^{-1}_{p,c} S_T c_0 (1-c_0)(\tau - \tau_{th})^{-1} \left[\tau(1-e^{-t/\tau}) - \tau_{th}(1-e^{-t/\tau_{th}})\right] \; . \tag{26}$$

Equation (26) contains a fast response from the temperature and a slow one from the concentration grating, whose time constants differ by roughly three orders of magnitude for typical polymer solutions and, hence, can easily be separated.

3.3 Measurement of Simple Liquid Mixtures

Equation (26) can directly be employed for the description of TDFRS experiments on simple liquid mixtures, which are characterized by a single diffusion time τ.

In Fig. 9, measurements on toluene/n-hexane and ethanol/water mixtures are shown. The symmetric response of the heterodyne signal after switching off the grating is obvious, and from the scaled-up residues it can be seen that there is only very little systematic deviation from the model curve. Due to the negative Soret-coefficient of the ethanol/water system, the diffraction efficiency shows a maximum shortly after the switching, when the temperature grating has already reached its steady state but the concentration grating has not yet developed. The numbers extracted from the measurements in Fig. 9 are D=3.38 (0.42) $\times$ 10^{-5} cm^2s^{-1} and S_T=4.54 (–3.25)$\times$ 10^{-3} K^{-1} for toluene/n-hexane (ethanol/water), respectively [49]. The ethanol/water result is in excellent agreement with values reported by Kolodner et al. [15]. For the toluene/n-hexane mixture, there has been considerable disagreement with literature data, which were obtained from different experimental techniques such as small-angle Rayleigh scattering [50] and thermogravitational column [5] measurements. In the meantime, these ambiguities have been resolved in favour of the TDFRS results [16, 51]. The discrepancies encountered for toluene/n-hexane exemplify the difficult and unsatisfactory experimental situation even for presumably simple systems.

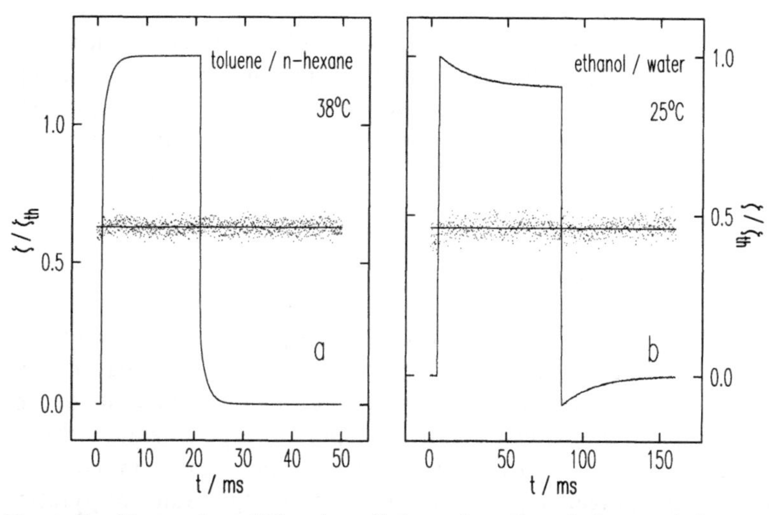

Fig. 9. Normalized heterodyne diffraction efficiency for toluene/n-hexane (toluene weight fraction 0.517) and ethanol/water (ethanol weight fraction 0.391). From Ref. [49]

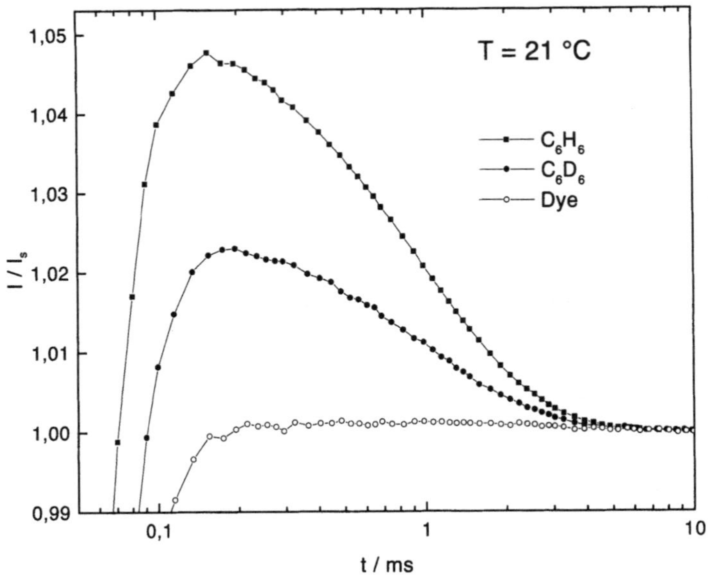

Fig. 10. Isotope effect: Normalized concentration mode for mixtures of protonated (C_6H_6) and deuterated (C_6D_6) benzene with cyclohexane (C_6H_{12}) at $c \approx 0.5$. Also shown is the signal for pure dyed cyclohexane

Figure 10 shows the concentration modes for two other liquid mixtures, namely cyclohexane (C_6H_{12}) with protonated (C_6H_6) and deuterated (C_6D_6) benzene, respectively, at approximately equal weight fractions, $c \approx 0.5$. Again, the signals have been normalized to the amplitude of the heat mode. Note the high quality of the data, where the concentration mode accounts for only a few percent of the entire diffracted signal.

Obviously, there is a strong isotope effect, and the Soret coefficient is reduced in the case of (C_6D_6), which has exactly the same mass as the other component, the cyclohexane. The measurements were conducted at T=21°C, $q = 8030\ cm^{-1}$, and the contrast factors are almost identical for both mixtures. The sign of S_T is such that benzene migrates towards the warmer regions [52].

3.4 Treatment of Polydispersity

All tools for polymer analysis must address the problem of polydispersity, and, generally, certain averages of the properties of interest are measured. The statistical weights with which the different molar masses contribute to the measured quantity are characteristic for the respective experimental technique and usually fixed. Well-known examples are the number average molar mass M_n as obtained from colligative properties or the z-average radius of gyration $\langle R_g \rangle_z$ from static light scattering. Narrow molar mass distributions are available only for a rather small number of synthetic polymers, and commodity polymers in particular show broad distributions spanning several decades in M. The wrong treatment and combination of different averages is a frequently encountered mistake.

In this section it will be outlined how the different molar masses contribute to the TDFRS signal. Of especial interest is the possibility of selective excitation and the preparation of different nonequilibrium states, which allows for a tuning of the relative statistical weights in the way a TDFRS experiment is conducted. Especially when compared to PCS, whose electric field autocorrelation function $g_1(t)$ strongly overestimates high molar mass contributions, a much more uniform contribution of the different molar masses to the heterodyne TDFRS diffraction efficiency $\zeta_{het}(t)$ is found. This will allow for the measurement of small molecules even when large amounts of high molar mass material are present.

The treatment of polydispersity becomes straight forward in the limit of low concentrations, where $c_0(1-c_0) \approx c_0 = \sum_k c_{0,k}$. $c_{0,k}$ is the concentration of component k with molar mass M_k. Under this assumption, Eq. (18) can be solved for every component independently by

$$c_k(x,t) = c_{0,k} + c_{q,k}(t)e^{iqx} \tag{27}$$

with

$$c_{q,k}(t) = -q^2 D_{T,k} c_{0,k} \int_{-\infty}^{t} dt' T_q(t') e^{-(t-t')/\tau_k} \quad . \tag{28}$$

Here, $D_{T,k}$ are the thermal and $D_k = (\tau_k\, q^2)^{-1}$ the translational diffusion coefficient of the k·th component, respectively. The diffraction efficiency $\zeta_{het}(t)$ is obtained according to Eq. (24) with a redefined memory function for $t \gtrsim 0$:

$$g(t) = \delta(t) + \sum_k a_k e^{-t/\tau_k} \tag{29}$$

$$a_k = -q^2 \left(\frac{\partial n}{\partial T}\right)_{p,c}^{-1} \left(\frac{\partial n}{\partial c_k}\right)_{p,T,c_{l \neq k}} D_{T,k} c_{0,k} \tag{30}$$

All potentially molar-mass-dependent quantities have been labelled with a summation index k in Eq. (30), which, in this form, also holds for dilute solutions of mixtures of chemically different species or copolymers with heterogeneity of both chemical composition and degree of polymerization.

In analogy to Eq. (26), the response of a polydisperse sample to an excitation step from 0 to 1 at $t = 0$ is given by

$$\zeta_{het}(t) = 1 - e^{-t/\tau_{th}} + \sum_k a_k \frac{\tau_k}{\tau_k - \tau_{th}} \left[\tau_k\left(1 - e^{-t/\tau_k}\right) - \tau_{th}\left(1 - e^{-t/\tau_{th}}\right)\right]. \tag{31}$$

For slowly diffusing polymer molecules with $\tau_k \gg \tau_{th}$, the response of the concentration mode in $\zeta_{het}(t)$ to an excitation pulse starting at $t = -\tau_p$ and ending at $t = 0$ is approximately given by

$$\zeta_{het}(t) = \sum_k a_k \tau_k (1 - e^{-\tau_p/\tau_k}) e^{-t/\tau_k} \quad . \tag{32}$$

This is simply a multiexponential decay with the amplitude of the k-th mode given by $a_k \, \tau_k \, (1 - e^{-\tau_p/\tau_k})$. In the short and in the long exposure limit, this reduces to

$$\zeta_{het}(t) = \begin{cases} \sum_k a_k \tau_p e^{-t/\tau_k} & \tau_p \ll \tau_k \\ \sum_k a_k \tau_k e^{-t/\tau_k} & \tau_p \gg \tau_k \end{cases} \quad . \tag{33}$$

Equation (33) can be further simplified in the case of homopolymers or copolymers with constant chemical composition. It is well known from the literature that D_T is independent of molar mass [11, 21, 36, 37], and the same holds for $(\partial n / \partial c_k)_{P,T,c_{l \neq k}}$ if M is sufficiently high to neglect end-group effects. Under these preliminaries, the molar mass dependence of a_k reduces to the concentration of the respective species: $a_k \propto c_{0,k}$.

Hence, since $q^2 \tau_k = D_k^{-1}$, the amplitude factors in the multiexponential $\zeta_{het}(t)$ are strictly proportional to the concentrations $c_{0,k}$ in the case of short exposures, where the memory function $g(t)$ is measured directly. In the long exposure limit, the amplitude factors are proportional to $c_{0,k} / D_k$.

From Eq. 33, it can be shown that the weight average diffusion coefficient

$$\langle D \rangle_c = \frac{\sum_k c_{0,k} D_k}{\sum_k c_{0,k}} \tag{34}$$

is obtained from a short exposure TDFRS experiment, whereas a long exposure experiment yields the average

$$\langle D \rangle_{c/D} = \frac{\sum_k c_{0,k}}{\sum_k c_{0,k} / D_k} \quad . \tag{35}$$

The molar-mass-independent D_T can be obtained from the initial slope of the concentration mode, whose steady state amplitude yields the weight average Soret coefficient [37]

$$\langle S_T \rangle_c = \frac{\sum_k c_{0,k} (D_{T,k} / D_k)}{\sum_k c_{0,k}} \quad . \tag{36}$$

The averages of the diffusion coefficient are calculated from the respective underlying rate distribution, which may be obtained from, e.g., a CONTIN analysis [53, 54]. For narrow distributions, a cumulant expansion [55] may be employed to obtain the average rate $\langle \Gamma \rangle$ and higher cumulants from a nonlinear least-squares fit. Interestingly, the average diffusion coefficient from a long exposure experiment yields directly the weight average hydrodynamic radius according to the Stokes-Einstein relation [37]:

$$\langle D \rangle_{c/D} = \frac{kT}{6\pi\eta_0 \langle R_h \rangle_c} \tag{37}$$

3.5 TDFRS Versus PCS

Naturally, all experimental techniques that allow the measurement of diffusion coefficients must be compared to PCS, which is the most widely employed method for the study of polymer diffusion processes.

The normalized electric field autocorrelation function $g_1(t)$, which can be calculated from the normalized intensity autocorrelation function $g_2(t) = \langle I(0)\, I(t) \rangle \langle I(0) \rangle^{-2}$ according to the Siegert relation [56]

$$g_1(t) = \left[\frac{1}{f}(g_2(t) - 1) \right]^{1/2} \tag{38}$$

serves as a starting point. $f \leq 1$ is a function of the number of coherence areas. $g_2(t)$ is measured in a typical homodyne experiment.

$g_1(t)$, as obtained from PCS, and $\zeta_{het}(t)$, as obtained from TDFRS, are in principle very similar. When measured for a monodisperse system at the same q-value, both show the same decay time. PCS utilizes the scattering from thermodynamic concentration fluctuations, and a q-vector is selected by the angle of observation. In TDFRS, on the other hand, a coherent excitation with a single well-defined q-vector is induced, whose decay is then observed. As a consequence, there are very different statistical weights for different molar masses in PCS and TDFRS, and even for TDFRS alone the relative statistical weights depend on the excitation time pattern. Since, for dilute solutions, the scattering from concentration fluctuations is proportional to the product of concentration times molar mass, cM, PCS shows an increased sensitivity for high molar masses, which strongly dominate $g_1(t)$ in case of polydisperse material. With TDFRS, the statistical weights are more uniform and, as a consequence, $\zeta_{het}(t)$ almost always decays faster than $g_1(t)$.

Before the relative statistical weights can be compared, it is necessary to relate the diffusion coefficient D to the molar mass M. Usually, this can be done by a scaling relation

$$\Gamma = q^2 D = aM^{-b} \,, \tag{39}$$

where the exponent b is characteristic for the fractal dimension of the polymer molecule or colloidal particle. $b=1/2$ for random coils under Θ-conditions, $b=1/3$ for solid spheres, and $b=1$ for thin rods.

It should be noted that Eq. (39) is slightly different from the usual definition, where $a\,M^{-b}$ is employed for D, not Γ. a, as defined in Eq. (39), is q-dependent because of the q^2.

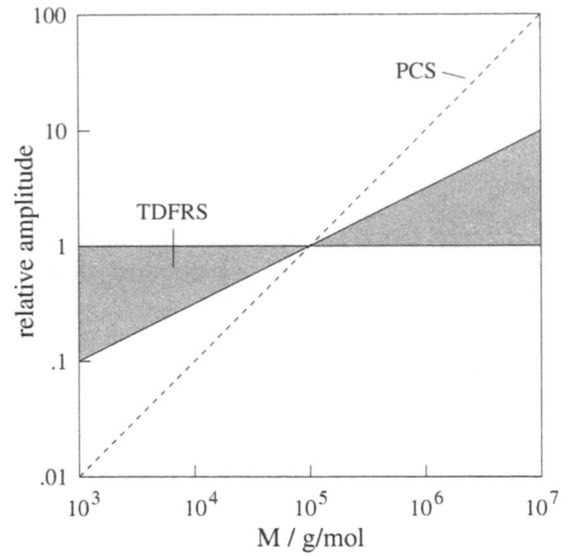

Fig. 11. Relative statistical weights for polydisperse polymers as a function of M. The line with slope 1 corresponds to PCS, the shaded area to TDFRS. The long exposure limit corresponds to the random coil value b=1/2

From Eq. (39), the amplitude factors in Eq. (33) can be expressed as a function of molar mass and concentration, and the normalized decay function for a dilute polymer solution, which is $\zeta_{het}(t)$ in case of TDFRS and $g_1(t)$ in case of PCS, can be written as

$$\sum_k p_k e^{-t/\tau_k} \quad \text{with} \quad p_\kappa = \frac{c_{0,k} M_k^\alpha}{\sum_\kappa c_{0,k} M_k^\alpha} \ . \tag{40}$$

The exponent α depends on the type of experiment performed, and it has been introduced to allow for a consistent description of the different measurements. α=1 for PCS, α=0 for short exposure TDFRS, which guarantees strictly concentration proportional amplitudes, and α=b for long exposure TDFRS. α = 1 leads directly to the z-average diffusion coefficient $\langle D \rangle_{cM}$ characteristic for PCS.

This situation is illustrated in Fig. 11, where the relative statistical weights and, hence, the amplitudes contributing to the multiexponential decays, are plotted as a function of M. The curves are arbitrarily normalized to unity at M = 10^5 g/mol. The strong underestimation of components with low molar mass in PCS is obvious. The shaded area is accessible to TDFRS, and the two limiting cases correspond to short and long exposure times.

Since the molar mass distribution of polydisperse polymers easily spans 3 to 4 decades, the low molar masses are masked by the strong scattering of the high-

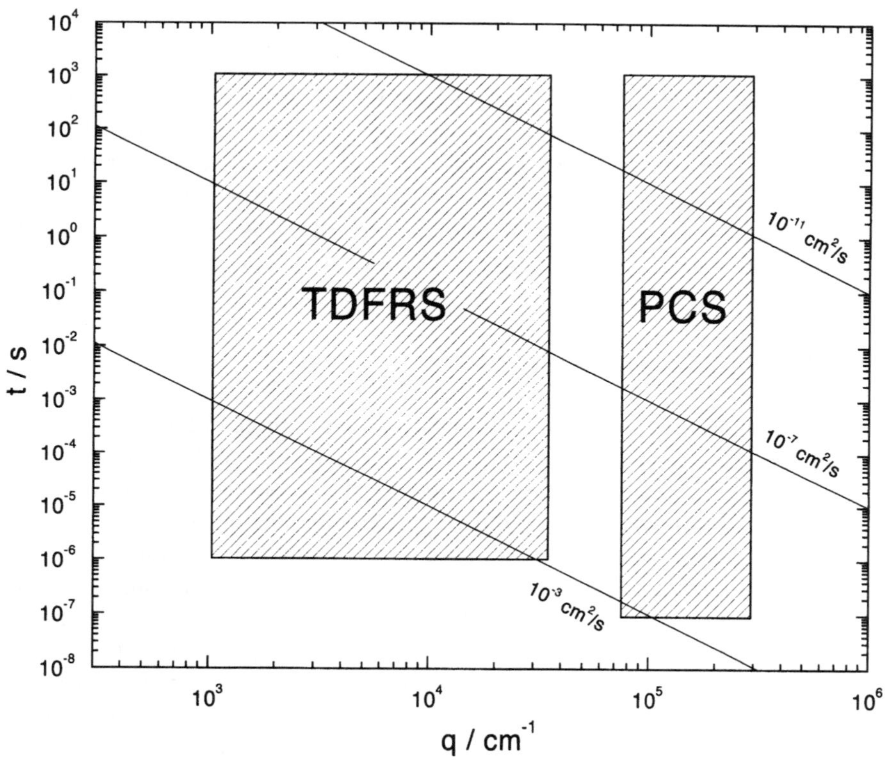

Fig. 12. Accessible q – t– range for typical PCS and TDFRS experiments. The lines correspond to diffusion coefficients $D = (q^2t)^{-1}$ of 10^{-3}, 10^{-7}, and 10^{-11} cm^2/s

er molar masses and, hence, are invisible in PCS, but not in TDFRS with its much more uniform amplitude distribution.

Accessible ***q–t*****-range:** Typical TDFRS experiments are performed at very low q-values, corresponding to scattering angles of only a few degrees, which eliminate the necessity for the extrapolation $q \to 0$ (hydrodynamic limit). PCS measurements, on the other hand, are usually performed at scattering angles of $30° < \theta < 150°$. The accessible time scales are comparable for both techniques and range from 1 μs or below to typically 10^3 s or longer. The lower limit is imposed by the electronics of the detection circuits and is, in principle, identical for both techniques.

Figure 12 gives an overview over the accessible q- and t-range for typical experiments. The numbers used are $0.5° < \theta < 15°$, $\lambda = 488$ nm (TDFRS) and $15° < \theta < 150°$, $\lambda = 647$ nm, n=1.5 (PCS). The wavelengths are the ones routinely employed in our laboratory. A change of wavelength does not significantly change Fig. 18. Note that

$$q = \frac{4\pi n}{\lambda} \sin\frac{\theta}{2} \tag{41}$$

in the case of cylindrical cells as usually used for PCS, whereas q is independent of n for cells with plane-parallel windows as employed for TDFRS [Eq. (10)].

The boundaries in Fig. 12 are soft, and PCS can be extended to lower and TDFRS to higher q, however at the expense of increased experimental problems, like stray light from the cuvette in case of PCS and a loss of signal amplitude, which is proportional q^{-2}, in the case of TDFRS.

4
Application to Polymer Analysis

This section is concerned with the information that can be obtained by TDFRS about the molar mass and size distribution and the various molar mass averages.

Once the scaling relation of Eq. (39) is known, the molar mass distribution can, at least in principle, be obtained from a Laplace inversion of the multi-exponential decay function as defined in Eq. (40). At this point, the differences between PCS and TDFRS stem mainly from the different statistical weights and from the uniform noise level in heterodyne TDFRS, which does not suffer from the diverging baseline noise of homodyne PCS caused by the square root in Eq. (38).

In any case, the rate distribution function $P(\Gamma)$, or the discrete p_k, must first be determined from $\zeta_{het}(t)$ or $g_1(t)$. Since this Laplace inversion is highly unstable, it is necessary to impose additional constraints, and various algorithms have been proposed in the literature [53, 54, 56, 57, 58, 59]. Thorough discussions of these methods with respect to PCS data can be found in Ref. [60, 61, 62]. Here we will concentrate on the CONTIN program developed by Provencher [53, 54, 57].

4.1
Rate Distribution and Averages of a Bimodal Polymer

Rate constants: A mixture of two narrow PS ($M_w/M_n \approx 1.03$) with molar masses of $M_{w,1} = 48$ kg/mol and $M_{w,2} = 556$ kg/mol at equal concentrations ($c_{0,1} = c_{0,2} = 0.0029$), dissolved in ethylacetate, has been used as a simple model system for a broad molar mass distribution [37]. To investigate the influence of exposure time, an exposure series with $0.025\ s < \tau_p < 2\ s$ has been recorded, and the respective normalized heterodyne diffraction efficiencies are plotted in the insert of Fig. 13. The exposure ranges from a time that is significantly shorter than the diffusion time of the fast component to a time that is much longer than the diffusion time of the slow component.

The main part of Fig. 13 shows the average rate $\langle\Gamma\rangle$, which closely follows the theoretical curve with the asymptotic values $\langle\Gamma\rangle_c$ in the short and $\langle\Gamma\rangle_{c/\Gamma}$ in the long exposure limit. There are no free parameters for the theoretical curve since the individual diffusion rates were obtained from the independently determined scaling relation for the diffusion coefficient, $D = 2.76 \times 10^{-4}\ \mathrm{cm^2 s^{-1}} \times M^{0.525}$.

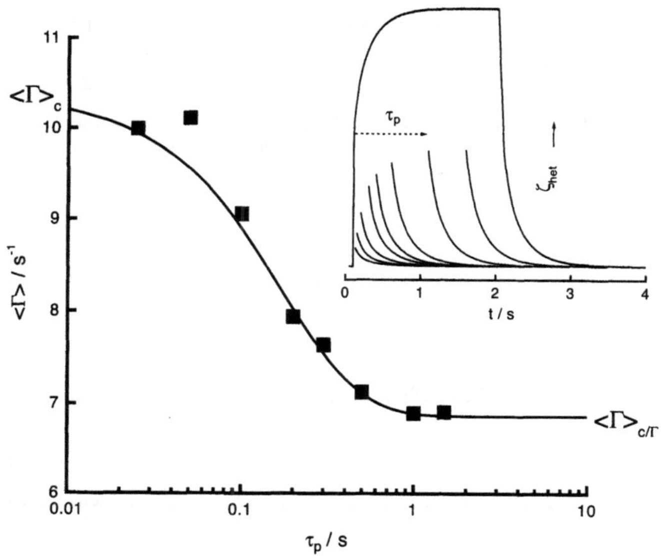

Fig. 13. Measured average rate as a function of exposure time τ_p. The solid line represents the theoretical prediction. The insert shows the normalized heterodyne signals from the exposure series. The fast temperature jumps after the grating is switched off at $t=\tau_p$ are omitted for clarity. From Ref. [37]

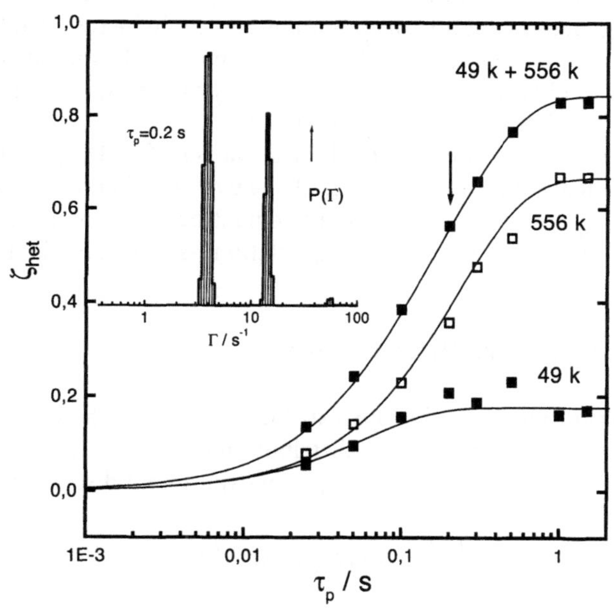

Fig. 14. Amplitudes of the normalized concentration signal and of the two individual modes as a function of exposure time. The insert shows the rate distribution for an exposure time $\tau_p = 0.2$ s. From Ref. [37]

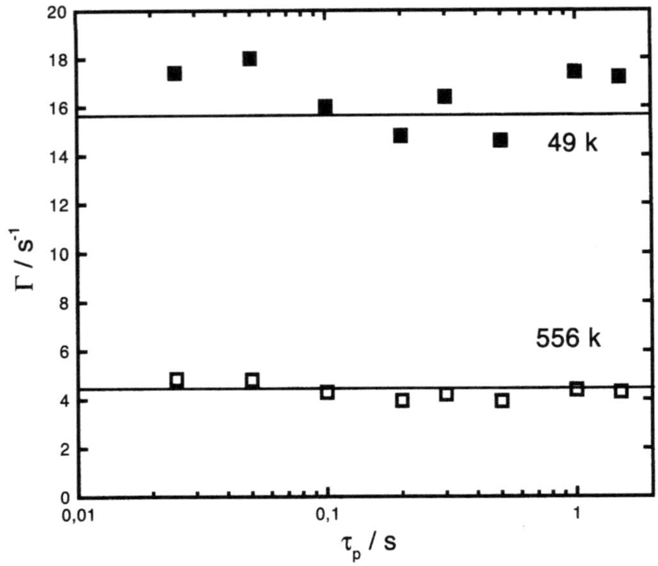

Fig. 15. Rates for the two components as a function of exposure time. The solid lines have been calculated from the scaling relation $D = 2.76 \times 10^{-4}\ \mathrm{cm^2 s^{-1}} \times M^{0.525}$

Fig. 14 shows the amplitudes of the two components and of the whole concentration signal as a function of τ_p. Both modes could be resolved for all exposure times, and the insert of Fig. 14 shows, as an example, the rate distribution for $\tau_p = 0.2$ s as obtained from a CONTIN analysis. The solid lines are the parameter-free predictions according to Eq. (32).

The two modes could be recovered for all values of τ_p. Despite the exposure time dependent average rate $\langle\Gamma\rangle$ in Fig. 13, the two rates Γ_1 and Γ_2 are almost independent of τ_p, and there is a good agreement between the measured values and the ones calculated from the scaling relation, as shown in Fig. 15.

4.2 Molar Mass Distribution and Molar Mass Averages

Bimodal distribution: Once the rate distribution $P(\Gamma)$ and the scaling relation in Eq. (39) are known, the molar mass distribution $c(M)$, expressed in weight fractions, is obtained for the three basic types of experiment discussed:

$$c(M) \propto P(\Gamma(M))M^{-\alpha}\left|\frac{d\Gamma(M)}{dM}\right| \propto P(\Gamma(M))M^{-1-b-\alpha} \quad . \tag{42}$$

Again, $\alpha = 1$ for PCS, $\alpha = 0$ for short exposure TDFRS, and $\alpha = b$ for long exposure TDFRS, and b is the scaling exponent from Eq. (39). For TDFRS with arbitrary exposure time, $c(M)$ is given by

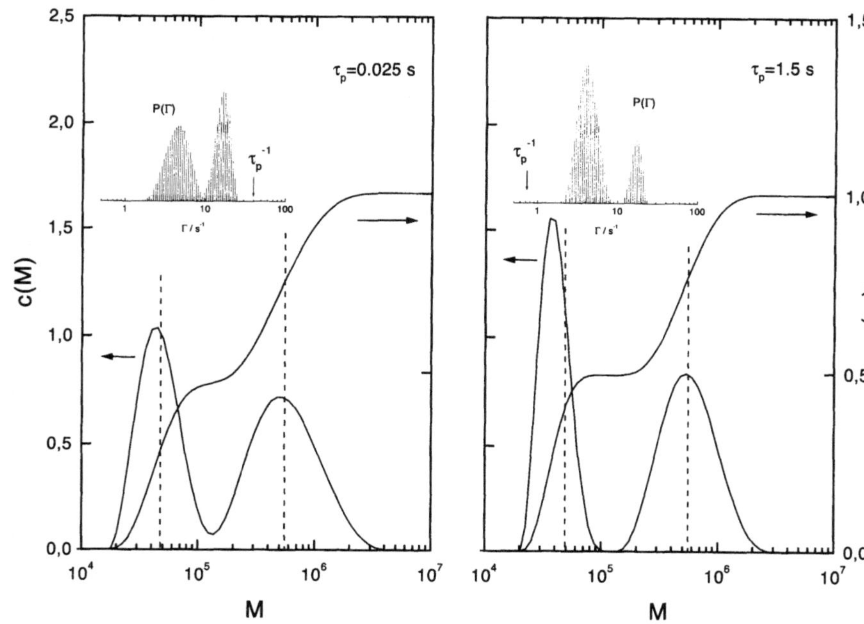

Fig. 16. Differential ($c(M)$) and cumulative ($I(M)$) molar mass distribution as calculated from $\zeta_{het}(t)$ for $\tau_p = 0.025$ s and $\tau_p = 1.5$ s. The inserts are the rate distributions as obtained from a CONTIN analysis. From Ref. [37]

$$c(M) \propto P(\Gamma(M))(1-e^{-\Gamma\tau_p})^{-1} M^{-1-2b} \quad . \tag{43}$$

Figure 16 shows the reconstruction of the molar mass distribution from two measurements in Fig. 13 according to Eq. (43) for an exposure time $\tau_p = 0.025$ s, which is shorter than the inverse of the fastest rate in the system, and for $\tau_p =$ 1.5 s, which is longer than the inverse of the slowest rate. The inserts show the respective rate distributions. It is evident that the slow mode is much more pronounced in $P(\Gamma)$ for the long exposure time. The peak positions in the differential molar mass distributions $c(M)$ agree reasonably well with the nominal molar masses, which are marked by the dashed vertical lines. The two peaks are somewhat broader and less resolved in the case of the shorter exposure. The peak areas in the molar mass distributions are, contrary to the rate distributions, identical for both measurements, as can be seen from the plateaus at $I(M) \approx 0.5$ in the cumulative molar mass distribution

$$I(M) = \int_0^M dM' c(M') \quad . \tag{44}$$

Broad distribution: A PS obtained by radical polymerization served as a model system with a broad smooth molar mass distribution. The solvent was ethylace-

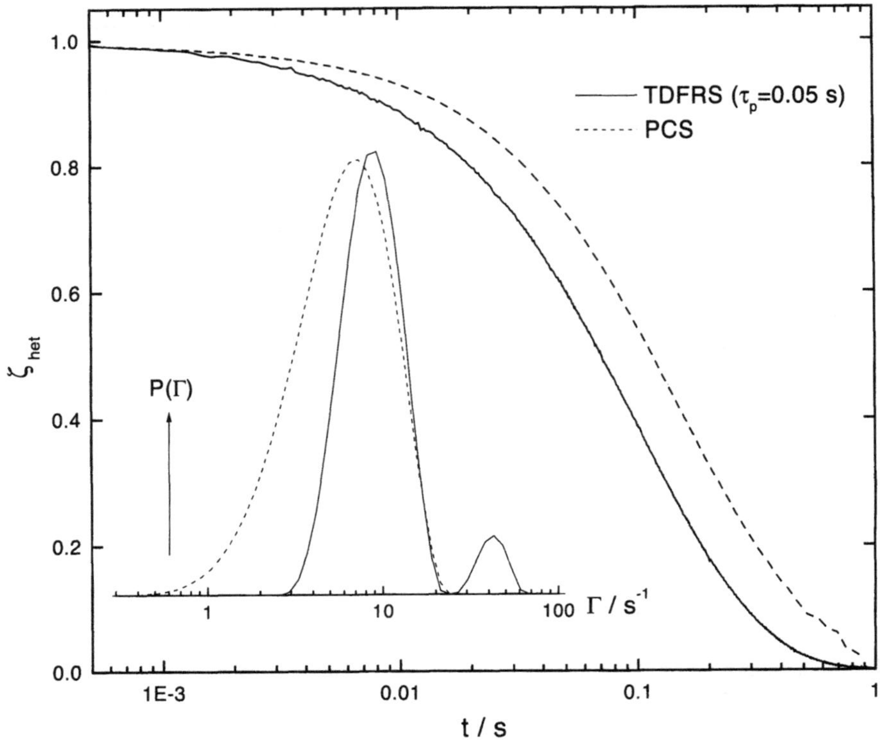

Fig. 17. Normalized decay curves from TDFRS and PCS, scaled to the same q-value, for a broad PS in ethyl-acetate. Rate distributions as inserts. From Ref. [37]

tate and the concentration $c_0 = 0.005$, and both PCS and TDFRS measurements with two different exposure times have been made [37]. $g_1(t)$ from PCS and $\zeta_{het}(t)$ from TDFRS with $\tau_p = 0.05$ s, approximately corresponding to short exposure conditions, are plotted together with the respective rate distributions in Fig. 17. $g_1(t)$ has been measured at $\theta = 90°$ and rescaled according to $\Gamma \propto q^2$ to the q-value of the TDFRS experiment, which was conducted at $\theta \approx 2°$. The slower decay of $g_1(t)$ is obvious.

Figure 18 shows the molar mass distributions as computed from the rate distributions according to Eq. (42) for the two measurements in Fig. 17 and from an additional long exposure TDFRS measurement with $\tau_p = 2$ s. Superimposed is the molar mass distribution obtained from size exclusion chromatography.

Obviously, none of the three scattering experiments is able to reproduce the true molar mass distribution exactly. The two TDFRS measurements show a bimodal structure, which must be regarded as an artifact. Otherwise, the general shape of the distribution, with the peak position around 200 kg/mol and a tail towards low M, is rather well reproduced. In principle unimodal distributions could be enforced by the choice of a larger regularization parameter. Purposely,

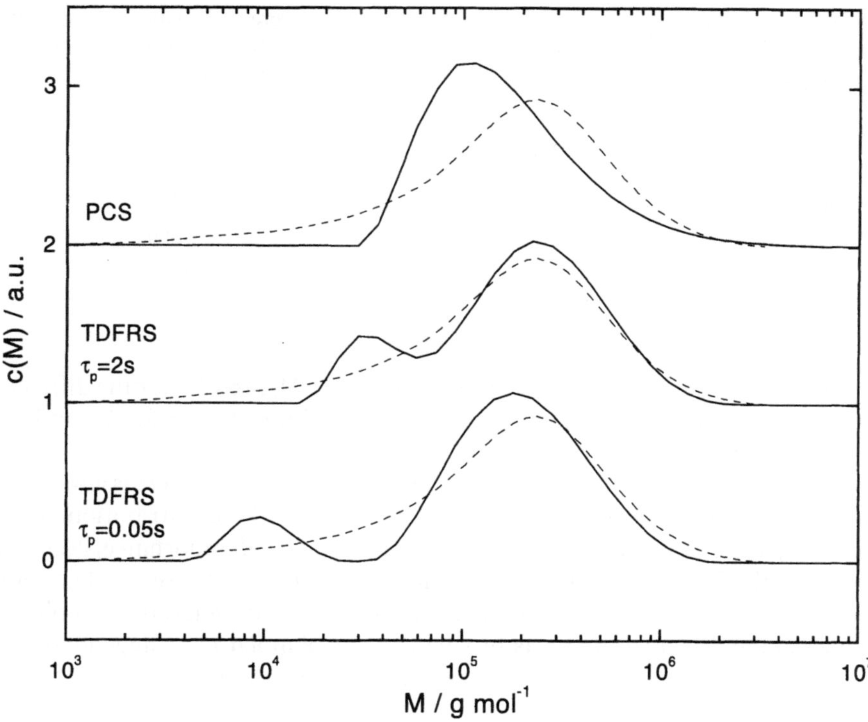

Fig. 18. Molar mass distributions calculated from PCS and TDFRS measurements with different exposure times. Superimposed is the molar mass distribution obtained from size exclusion chromatography. From Ref. [37]

this additional a-priori knowledge has not been utilized and the *chosen solution* has been taken. It has been found that multiple peaks are frequently present in rate distributions obtained by CONTIN. The lower moments of the distribution are, however, usually rather insensitive to this fine structure. The PCS measurement produces a unimodal peak, but there is substantial deviation in the wings of the distribution. Especially the low-M tail is completely missing as a consequence of the weak contribution of low molar masses to $g_1(t)$.

Frequently, not the fine structure of a distribution is of interest but certain molar mass averages, corresponding to the lower moments of the distribution, especially M_w and M_n. They can be computed directly from $P(\Gamma)$ and Eq. (39) according to

$$M_w \equiv \langle M \rangle_c = a^{1/b} \langle \Gamma^{(\alpha-1)/b} \rangle_{cM^\alpha} \langle \Gamma^{\alpha/b} \rangle_{cM^\alpha}^{-1} \tag{45}$$

$$M_n \equiv \langle M \rangle_{c/M} = a^{1/b} \langle \Gamma^{\alpha/b} \rangle_{cM^\alpha} \langle \Gamma^{(\alpha+1)/b} \rangle_{cM^\alpha}^{-1} \; . \tag{46}$$

Table 1. Number average (M_n), weight average (M_w), and z-average (M_z) molar masses (in kg mol^{-1}) of a broad PS as measured by different techniques. TDFRS (0.05 s) corresponds approximately to the short and TDFRS (2 s) to the long exposure limit

	M_n	M_w	M_z
SEC	57	285	640
TDFRS (0.05 s)	58	237	434
TDFRS (2 s)	107	275	509
PCS	121	262	1229

Table 1 summarizes the molar mass averages M_n, M_w, and M_z for the different experimental techniques. Good agreement is found for M_w, less for M_n and M_z with particularly strong deviations for the PCS results.

When such different techniques as in Table 1 are compared, there is always the problem of different sensitivities for different aspects of the distribution. If, for example, information about the high molar mass tail is of importance, PCS may be the method of choice. It may also be incorrect to regard the SEC distribution as the true molar mass distribution as it may suffer from calibration problems, solute-column interactions, peak broadening, and a molar mass dependence of the contrast factor $\partial n/\partial c$, and hence the detector sensitivity.

4.3 Weight Distribution of the Particle Size

From both scattering techniques, PCS and TDFRS, it is possible to determine the rate distribution $P(\Gamma)$, or the p_k in the discrete case. By means of the Stokes-Einstein relation,

$$\Gamma q^{-2} = \frac{kT}{6\pi\eta_0 R_h}, \tag{47}$$

$P(\Gamma)$ can be transformed into a distribution of the particle size as defined by the hydrodynamic radius R_h. But only for TDFRS, and not for PCS, a particle size distribution in terms of weight fractions can be obtained without any prior knowledge of the fractal dimension of the polymer molecule or colloid, which is expressed by the scaling relation of Eq. (39). This can be seen from the following simple arguments:

$P(\Gamma)$ is the rate distribution as obtained from $\zeta_{het}(t)$ or $g_1(t)$. From the continuous form of Eq. (40) it follows that

$$c'(\Gamma) \propto P(\Gamma)\Gamma^{\alpha/b} \tag{48}$$

where $c'(\Gamma)$ is the weight fraction of particles with diffusion rate Γ. The weight fraction $c''(R_h)$ of particles with hydrodynamic radius R_h is

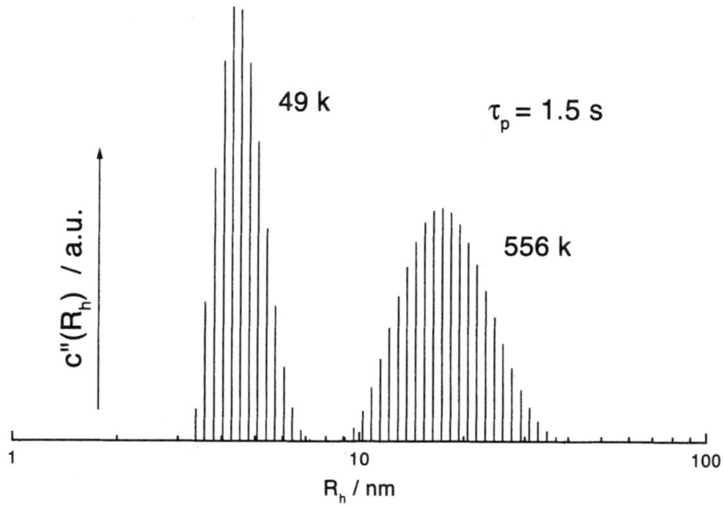

Fig. 19. Polymer weight fraction as a function of hydrodynamic radius R_h for the TDFRS measurement in Fig. 16 with $\tau_p = 1.5$ s

$$c''(R_h) = c'(\Gamma)\left|\frac{d\Gamma(R_h)}{dR_h}\right| \quad . \tag{49}$$

$\Gamma(R_h)$ is given by Eq. (47). From Eqs. (47) - (49) one obtains

$$c''(R_h) \propto P(\Gamma(R_h))R_h^{-2-\alpha/b} \quad . \tag{50}$$

For simplicity, normalization factors and proportionality constants have been omitted. Hence, for short and long exposure TDFRS, Eq. (50) reduces to $c''(R_h) \propto P(\Gamma(R_h))\, R_h^{-2}$ and $c''(R_h) \propto P(\Gamma(R_h))\, R_h^{-3}$, respectively. Eq. (50) can easily be generalized for TDFRS with arbitrary exposure times τ_p by means of Eq. (32). In the case of PCS, Eq. (50) becomes $c''(R_h) \propto P(\Gamma(R_h))\, R_h^{-2-1/b}$, and $c''(R_h)$cannot be obtained without knowing b. Fig. 19 shows $c''(R_h)$ as obtained from $P(\Gamma)$ from the insert in the right-hand half of Fig. 16 with $\tau_p = 1.5$ s.

4.4
Collective and Self-Diffusion of PS Microgels

TDFRS has been employed for the study of collective and self-diffusion of PS micronetwork spheres (microgels) [63] with crosslink ratios of 1:10, 1:20, and 1:50 in toluene from low to high concentrations.

Figure 20 shows ζ_{het} for a concentration series of microgels with a crosslink ratio of 1:20. From the long exposure measurements, the weight average hydrodynamic radius $\langle R_h \rangle_c = 19.0$ nm and a distribution width of

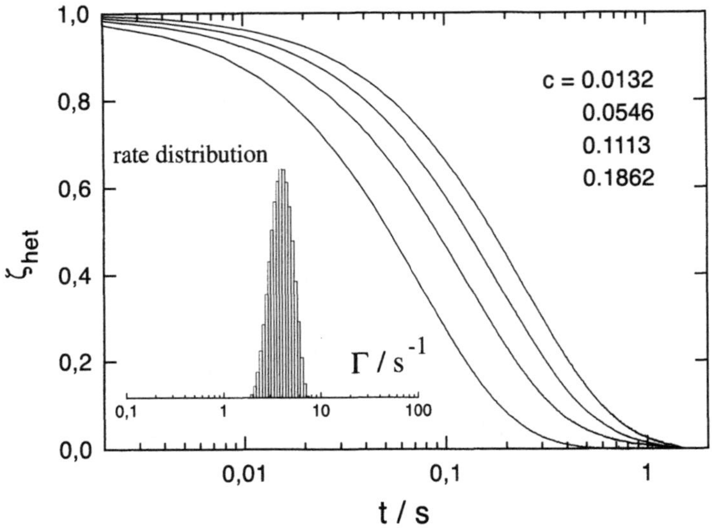

Fig. 20. Normalized heterodyne diffraction efficiencies for a concentration series of 1:20 crosslinked PS microgels in toluene. The insert shows the rate distribution for the lowest concentration, corresponding to the slowest decay curve. From Ref. [64]

$$[\langle R_h^2 \rangle_c / \langle R_h \rangle_c^2 - 1]^{1/2} = 0.27$$

is obtained [64]. Assuming solid spheres ($b = 1/3$, Eq. (39)), this translates to a polydispersity of $M_w / M_n = 1.7$, which is calculated from the rate distribution according to [37]

$$\frac{\langle \Gamma^2 \rangle_{cM^\alpha} - \langle \Gamma \rangle_{cM^\alpha}^2}{\langle \Gamma \rangle_{cM^\alpha}^2} = b^2 \left(\frac{M_w}{M_n} - 1 \right) . \tag{51}$$

The investigation of concentrated dispersions was motivated by the finding that microgels with a crosslink ratio of 1:50 show a significant deviation from hard-sphere behaviour [65]. From small angle neutron scattering (SANS), the deswelling of the micronetworks with increasing concentration was found to depend on the crosslink density [66]. FRS measurements with dye-labelled microgels at low and intermediate concentrations showed no influence of the crosslink density on the long-time self-diffusion coefficient D_s^L [67], which can be well described by the Medina-Noyola model for hard spheres [68].

The collective diffusion coefficient D, on the other hand, is very sensitive to the crosslink density, and TDFRS is well suited for its measurement since, because of the small scattering angles, there is no need to extrapolate for $q \rightarrow 0$.

Figure 21 shows the reduced diffusion coefficients D_s^L / D_0 and D/D_0 as a function of the volume fraction ϕ [67]. For the crosslink ratios 1:20 and 1:50 there is

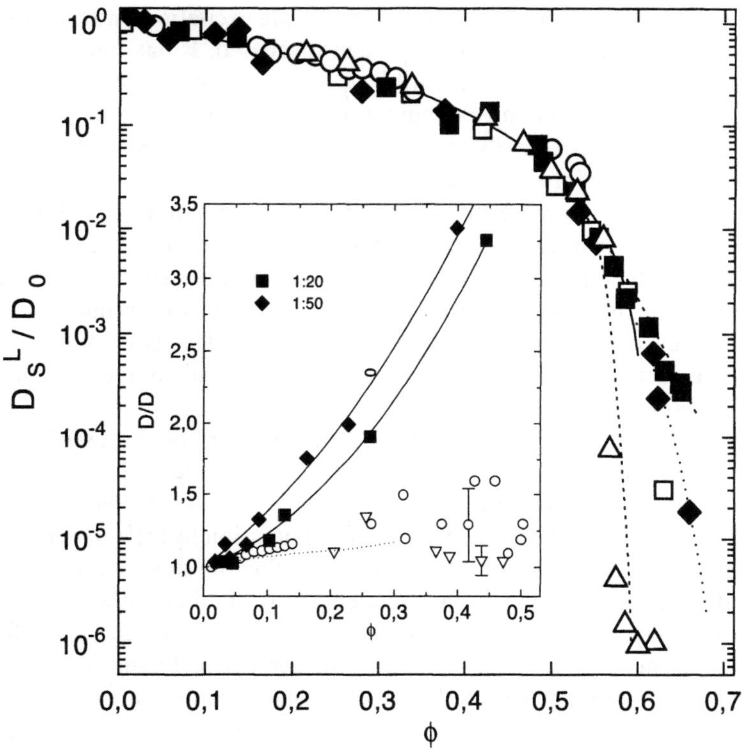

Fig. 21. Volume dependence of $D_s^L|D_0$ and D/D_0. Collective diffusion coefficients D: □ 1:20 and ◇ 1:50 (TDFRS). Hard-sphere results for comparison (○ [69,70], ▽ [71], and theoretical predictions (dashed line) [72]). From Ref. [67]

a pronounced deviation from the hard-sphere behaviour, which, as expected, increases with decreasing crosslink density.

For concentrations close to the glass transition, there is an additional slow mode observed both in PCS and TDFRS measurements, which might be explained by a coupled motion of large and small microgels on a short timescale. They decouple on a longer timescale, and their decoupled motion is essentially equivalent to microgel self-diffusion [64]. It becomes visible in TDFRS because of the different Soret-equilibria of the different-sized microgels. Interestingly, the amplitude of the slow mode in TDFRS is negative when compared to the fast mode. These results are, however, still very preliminary, and more experiments are needed to support the model.

5 Stochastic TDFRS

In the previous sections it has been shown that there are similarities between TDFRS and PCS, and that the main advantage of TDFRS stems from the possibility of tailored excitations. The two most simple excitation patterns, long and short excitation pulses, have already been discussed. It has been shown that short δ-excitation pulses, which satisfy $\tau_p \ll \tau_k$, are of particular importance since they allow the direct measurement of the memory function $g(t)$, to which the individual species contribute strictly in proportion to their concentrations and independently of their molar masses M_k.

Of course, there is a major drawback of short excitation pulses. Since, due to sample heating, the laser intensity cannot be increased in the same way as the pulses are shortened, the product $I\,\tau_p$ decreases, and, as a consequence, the signal vanishes in the baseline noise. When viewed from the frequency domain, this is a direct consequence of the low spectral power density of short excitation pulses. The decrease of the concentration signal is already obvious for the short excitation times in Fig. 13.

In this section, more general excitation patterns will be discussed, which allow for tailored deconvolutable excitations with high spectral power densities. Periodic amplitude modulation with a single frequency has been proposed in the literature [73]. It is well suited for noise suppression by lock-in detection, but in practice it suffers from stability problems during the slow frequency sweep.

Pseudostochastic random binary sequences in combination with fast Fourier transform and correlation techniques avoid these problems and allow for a direct measurement of $g(t)$ with high spectral power density and frequency multiplexing (stochastic TDFRS). Tailoring of the pseudostochastic sequences even allows for a selective enhancement and suppression of certain frequencies and, hence, of certain molecular species [74].

Pseudostochastic random binary sequences are noise-like time patterns. They are defined at times $n\,\Delta\,t$ and assume only two different values, corresponding to the grating amplitudes -1 and $+1$, if 180°-phase modulation is used for switching off the optical grating. Only software modifications, and no changes in the hardware of the TDFRS setup, are necessary in order to utilize pseudostochastic excitation sequences. The timing for heterodyne/homodyne separation is identical to the one already described for pulsed excitation.

In the following, a general treatment of arbitrary binary excitation sequences will be given. Since the proper definition of the excitation and the response function is not unambiguously possible, a problem-independent notation will first be given, which will later be mapped to the actual experiment. For the moment, it is sufficient to picture a linear system with an input $x(t)$, an output $y(t)$ and a linear response function $h(t)$, as sketched in Fig. 22. The input $x(t)$ may be a pulse of finite duration, as discussed in the previous sections, or a pseudostochastic random binary sequence as in Fig. 22.

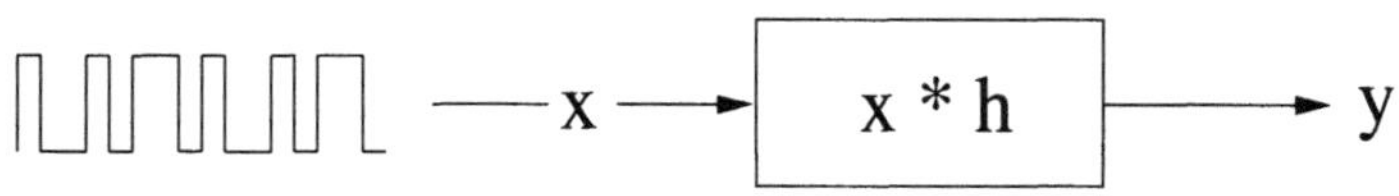

Fig. 22. Idealized picture of the experiment as a linear system characterized by a linear response function $h(t)$. The asterisk stands for convolution

5.1 Time and Frequency Domain Representation

Notation and basic equations: Following the representation in [75], the notation and the basic equations are as follows.

Continuous functions and signals in the time domain are denoted by lower case letters with the argument in parentheses, e.g. $x(t)$. Sampling at constant intervals Δt produces a discrete approximation $x[n]$ to the continuous signal, defined at times $t = n\,\Delta t$, $n = 0,1,2 \ldots$. Square brackets are used for the arguments of discrete functions. The Fourier transform establishes the connection between the time and frequency domains [76]:

$$x(t) = \frac{1}{2\pi} \int_{-\infty}^{\infty} d\omega\, X(\omega)\, e^{i\omega t} \tag{52}$$

$$X(\omega) = \int_{-\infty}^{\infty} dt\, x(t)\, e^{-i\omega t} \tag{53}$$

Frequency domain functions are denoted by upper case letters. Of importance for the TDFRS experiment is the discrete Fourier transform of an array of N data points within a period of $N\,\Delta t$:

$$x[n] = \frac{1}{N} \sum_{\kappa=0}^{N-1} X[k] e^{2\pi\, ink/N} \tag{54}$$

$$X[k] = \frac{1}{N} \sum_{n=0}^{N-1} x[n] e^{-2\pi\, ink/N} \tag{55}$$

The discrete and continuous frequency scales are related by

$$\omega_k = \frac{2\pi k}{N\Delta t} \,. \tag{56}$$

The response $y(t)$ of a linear system to an excitation $x(t)$ is a convolution of $x(t)$ with the response function $h(t)$. The TDFRS experiments are performed with periodic boundary conditions and discrete sampling, asking for the discrete periodic convolution

$$y[n] = x[n] * h[n] \equiv \sum_{i=0}^{N-1} x[i] h[(n-i) \bmod N] \,. \tag{57}$$

The convolution theorem reduces Eq. (57) to a simple product in frequency space:

$$Y = XH \tag{58}$$

The discrete periodic correlation is defined in analogy as

$$y[n] = x[n] \times h[n] \equiv \sum_{i=0}^{N-1} x[i]h[(n+i) \bmod N] , \quad (59)$$

$$Y = X^* H . \quad (60)$$

X^* is the complex conjugate of X. The arguments are omitted in Eqs. (58) and (60), which apply to both the continuous and the discrete case.

According to Eqs. (58) and (60), h is obtained from the deconvolution

$$H = \frac{Y}{X} = \frac{YX^*}{|X|^2} . \quad (61)$$

The transition to the time domain is established by Eq. (52). From the right-hand side of Eq. (61), it follows that h may also be determined by cross correlation of the sample response y with the excitation x in the case of a white power spectrum of the excitation with $|X|^2 = 1$.

5.2 Periodic TDFRS

Before stochastic TDFRS is treated in detail, periodic amplitude modulation of the grating in combination with phase-sensitive lock-in detection, similar to the procedure proposed by Bloisi [73], will be briefly discussed. With periodic amplitude modulation with a single frequency, which is slowly scanned through the frequency range of interest, the Fourier transform of the TDFRS response function, $G(\omega)$, is measured more or less directly.

If the retardation of the temperature grating is neglected for the moment, which is valid if $\Delta t \gg \tau_{th}$, a sinusoidal amplitude modulation of the optical grating translates directly into a sinusoidal amplitude modulation of the temperature grating with amplitude $\delta T = \tau_{th}\, \alpha I_0\, (\rho c_p)^{-1}$ according to Eq. (16):

$$T_q(t) = \delta T \, e^{i\omega t} . \quad (62)$$

The heterodyne diffraction efficiency [Eq. (24)]has the same periodicity:

$$\zeta_{het}(t) = G(\omega)\, e^{i\omega t} . \quad (63)$$

The complex $G(\omega)$ is the Fourier transform of $g(t)$ [Eq. (24)] according to Eq. (53), which, for polydisperse samples, is a superimposition of Lorentzians:

$$G(\omega) = G'(\omega) - iG''(\omega) = 1 + \sum_k a_k \frac{\tau_k}{1 + i\omega\tau_k} . \quad (64)$$

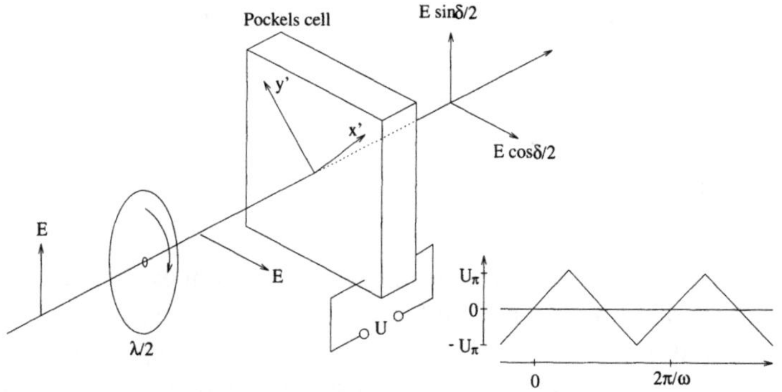

Fig. 23. Setup for periodic amplitude modulation. x' and y' are the two axes of the Pockels cell. They are rotated by 45° with respect to the polarization of the laser beam. $\delta = U/U_\pi$, and U_π is the voltage for 180° phase shift.

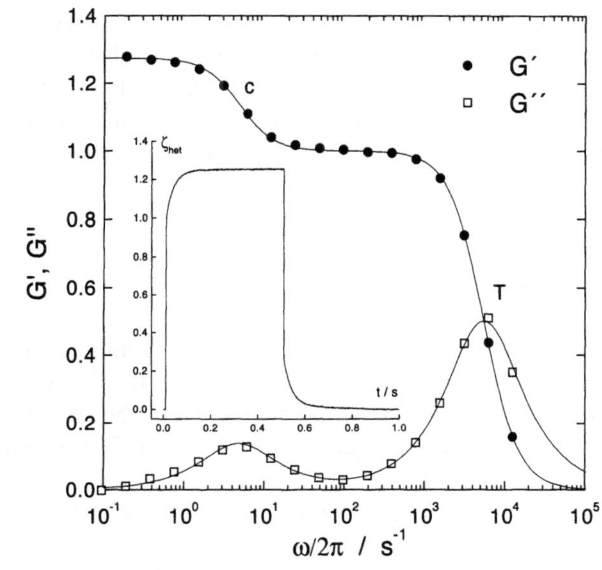

Fig. 24. Real and imaginary part of $G(\omega)$ as measured directly in a frequency domain experiment. The insert shows the corresponding measurement in the time domain (PS, M=49 kg/mol in toluene)

From an experimental point of view, a sinusoidal amplitude modulation of the optical grating can be achieved with the setup shown in Fig. 23 in one of the writing beams.

The real and imaginary part of $G(\omega)$ Fig. 24 has been measured in for an almost monodisperse PS (M = 49 kg/mol) in toluene. The slow mode at $\omega/2\pi \approx$ 5 Hz corresponds to mass diffusion. The fast mode at $\omega/2\pi \approx 5$ kHz stems from

Table 2. Diffusion time τ and amplitude of the concentration signal ζ_{het} as measured by the time and the frequency domain experiment in Fig. 24

	τ / ms	ζ_{het}
Time domain	32.4	0.265
Frequency domain	33.2	0.274

the retardation of the temperature grating, which was not explicitely considered in Eq. (64). The solid lines correspond to two Debye processes.

The corresponding time domain experiment with a long exposure pulse is shown in the insert. Both measurements have been normalized to the amplitude of the signal from the temperature grating. The amplitudes of the concentration signal and the diffusion time constant τ agree between both experiments within the experimental error (Table 2).

The main problem with periodic TDFRS is, that the different frequencies are measured at different times. This requires a long-time stability, especially of the heterodyne reference, lasting about as long as the entire experiment. Time domain experiments, on the other hand, are frequency multiplexed, and stability of the heterodyne background is only required for one homodyne/heterodyne separation cycle as described in the experimental section, which is only of the order of seconds, not hours. No stability of the signal amplitude is required for the averaging of $\zeta_{het}(t)$ over arbitrary times.

5.3 Identification of the Response Function

Returning to pseudostochastic excitation in the time domain. Figure 22 gives an idealized view of a real experiment. For stochastic TDFRS, the switching process of the grating becomes increasingly important and must be analyzed in more detail. A realistic picture of signal generation in a real TDFRS experiments is shown in Fig. 25.

Start and end point of the signal flow is the computer, where the excitation pattern is generated and the sample response collected.

A discrete binary sequence $x[n]$ of 1 and −1, defined at times $n\,\Delta t$, $n = 0,1,2 \ldots N-1$, is converted to a continuous binary voltage signal $x(t)$, which drives the high voltage generator for the Pockels cell. The resulting amplitude of the optical interference grating, $x'(t) \equiv I_q(t)/I_0$, excites the sample, which in principle could be described by a single response function $h'(t) = g_T(t) * g(t)$. $h'(t)$ is the convolution of the heat and the concentration modes as indicated by the dashed bracket in Fig. 25.

Because of the time scale separation of 3 to 4 decades between both modes, it is of advantage to convolute the heat mode $g_T(t) = \tau_{th}^{-1}\, exp(-t/\tau_{th})$ into the effective excitation:

$$x''(t) = T_q(t)\left[\frac{\alpha}{\rho c_p} I_0 \tau_{th}\right]^{-1} \tag{65}$$

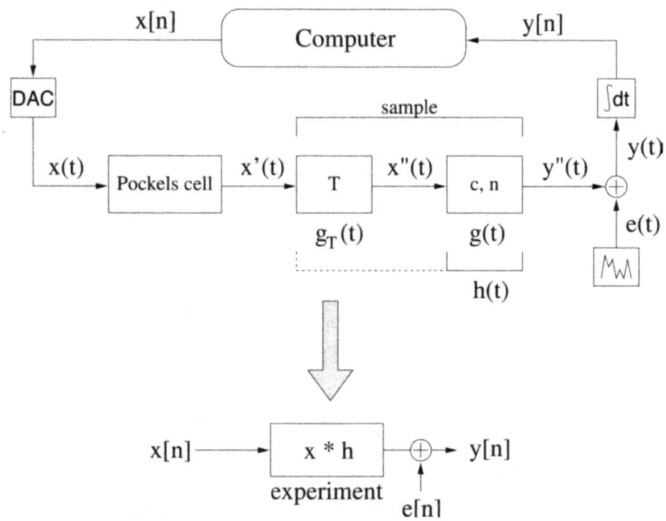

Fig. 25. Excitation and signal flow in a TDFRS experiment. The lower part shows the idealized picture used for data evaluation. From Ref. [75]

$T_q(t)$ is defined in Eq. (16). In the limit $\Delta t \gg \tau_{th}$, $x''(t) \approx x(t)$, except for a constant amplitude factor.

Now, the effective linear response function $h(t)$ can be identified with $g(t)$ as defined in Eqs. (25) and (29): $h(t) \equiv g(t)$. The primary sample response is the heterodyne diffraction efficiency $y''(t) \equiv \zeta_{het}(t)$. The instantaneous contribution of the temperature grating to the diffraction efficiency is expressed by the δ-function in $g(t)$ [Eq. (25)]. After the sample, an unavoidable noise term $e(t)$ is added. The continuous $y(t)$ is sampled by integrating with an ideal detector over time intervals Δt to finally obtain the time-discrete sequence $y[n]$.

Hence, the TDFRS experiment converts an ideal time-discrete excitation $x[n]$ into a time discrete signal $y[n]$, and the task is to extract the response $h(t) \equiv g(t)$ from the measured $y[n]$ and the known $x[n]$ according to Eq. (61).

In Ref. [75], it is discussed in more detail why it is advantageous to convolute the response of the temperature grating into the excitation and how to treat systematic errors arising from this approximation and from imperfections of the components in the setup. Especially the switching properties of the Pockels cell require careful analysis, since the switching number increases from 2 in case of pulsed excitation to approximately N in case of pseudostochastic binary sequences.

Figure 26 shows a stochastic TDFRS measurement. The system is PS ($M = 250$ kg/mol, $c = 0.0205$) in toluene. The upper half of the insert shows a short sequence of the pseudostochastic binary excitation sequence of length 2^{11} [75]. The sampling time Δt is 140 μs. How the sequence has been generated will be discussed in the next section.

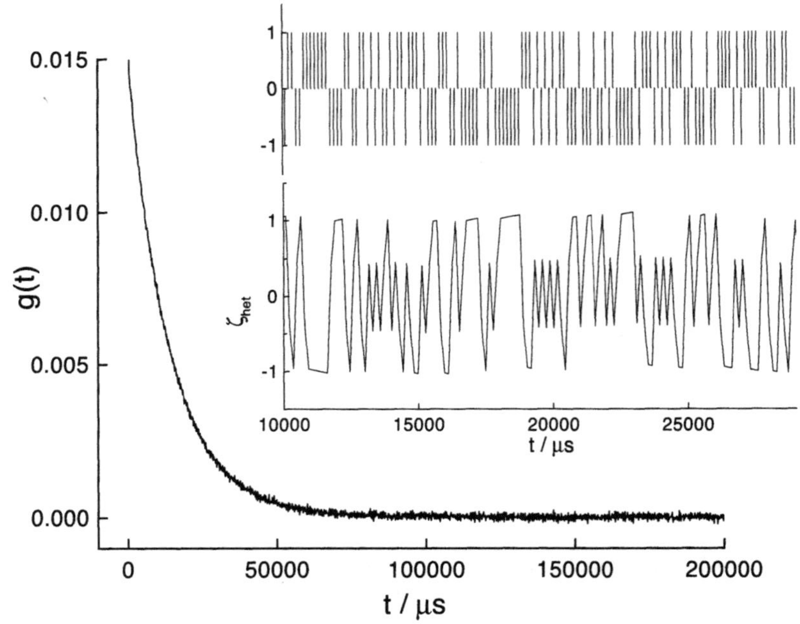

Fig. 26. Memory function $g(t)$ as obtained from stochastic TDFRS with an optimized random binary sequence. The inserts show a short sequence of the excitation and the corresponding heterodyne response. From Ref. [75]

The lower half of the insert shows the heterodyne diffraction efficiency as seen by the detector. It has a random character, but the influence of the memory function is obvious when compared to the excitation. The main part of Fig. 26 shows the concentration part of the memory function $g(t) \approx h(t)$ after deconvolution according to Eq. (61). The amplitude of the contribution from the temperature grating is normalized to unity and contributes only to the very first data point.

5.4 Criteria for Pseudostochastic Excitation Sequences

Not every arbitrary random binary sequence is well suited. In the following, design criteria for sequences that serve particular requirements will be given.

Sampling time: The timing of the measurements is controlled by a hardware clock on a computer plug-in board, whose clock rate is set to Δt^{-1}. At every tick of the clock, one data point is collected in such a way that the PMT counts accumulated between $t = n\,\Delta t$ and $t = (n+1)\,\Delta t$ are stored in $y[n]$. Similarly, $x[n]$ is applied to the output at $t = n\,\Delta t$, which keeps the value until $t = (n+1)\,\Delta t$.

The sampling time Δt must be sufficiently short to avoid aliasing from signal intensity at frequencies above the Nyquist frequency $\omega_{Ny} = \pi/\Delta t$, which are folded back into the frequency interval $-\omega_{Ny} < \omega_k < \omega_{Ny}$. As a rule, $\Delta t \lesssim \tau_{min}/10$ should be fulfilled, where τ_{min} is the shortest diffusion time constant.

If the finite thermal rise time τ_{th} cannot be neglected, which means that $\Delta t \gg \tau_{th}$ is not fulfilled, it may be necessary to sample at a higher rate than the clock rate of the excitation sequence (oversampling) [74]. A detailed discussion of this effect is, however, beyond the scope of this article.

Noise amplification: Malformed sequences are those with holes in the power spectrum, since the noise term *E[k]* is amplified at frequencies with low power density in the excitation. Quantitatively, this can be described as follows.

The total power P in the excitation is obtained from Parseval's theorem:

$$P = \sum_{k=0}^{N-1} |X[k]|^2 = N \sum_{n=0}^{N-1} x[n]^2 \tag{66}$$

Since the power at zero frequency, $|X[0]|^2$, merely adds a constant background without much information content, the requirement for maximum excitation power P_+ at positive frequencies is

$$P_+ = \sum_{k=1}^{N-1} |X[k]|^2 = N \sum_{n=0}^{N-1} x[n]^2 - \left(\sum_{n=0}^{N-1} x[n] \right)^2 = N^2 \left(\langle x^2 \rangle - \langle x \rangle^2 \right), \tag{67}$$

which is maximized for $\langle x \rangle = 0$. Hence, an excitation sequence should be approximately symmetric with respect to +1 and −1.

Mathematically, the deconvolution according to Eq.(61) can be carried out as long as $X[k] \neq 0$. In practice, there is always experimental noise $e[n]$, or $E[k]$ in the frequency domain, which is amplified at frequencies where $|X[k]|^2$ is small.

When the noise term is included, the deconvolution of Eq. (61) yields the unperturbed response function H plus a perturbation $E\, X^* |X|^{-2}$:

$$\frac{YX^*}{|X|^2} = H + E \frac{X^*}{|X|^2} \; . \tag{68}$$

From Eq. (68), the amplification U_k of the spectral power density of the noise is

$$U_k = \left| \frac{X^*[k]}{|X[k]|^2} \right|^2 = |X[k]|^{-2} \; , \tag{69}$$

and the integral amplification of white noise is defined as

$$U = \sum_{k=0}^{N-1} |X[k]|^{-2} \; . \tag{70}$$

The minimum $U = 1$ is assumed for a constant white power spectrum with $|X[k]|^2 = N$. All deviations from the constant power spectrum increase the integral noise amplification U. They may, however, be useful for noise suppression at selected frequencies. Arbitrary random binary sequences usually have a broad power spectrum with a high noise amplification due to a large number of holes with almost vanishing spectral power density.

The signal-to-noise ratio (SNR) corresponding to Eq. (68) is

$$SNR = \frac{\langle |H|^2 \rangle}{\langle |E|^2 |X|^{-2} \rangle} \quad . \tag{71}$$

Note that Eq. (71) is definied as a power ratio, contrary to the amplitude ratio in Eq. (4).

5.5 Maximum Length Binary Sequences

Maximum length binary sequences (MLBSs) of length $N = 2^L-1$, where L is a positive integer, have a perfectly flat power spectrum [77]. The deconvolution in Eq. (61) can be computed very efficiently by means of a fast Hadamard transform, and they have, for example, been employed for Hadamard NMR spectroscopy [78].

MLBSs are generated by a shift register of length L with feedback as shown in Fig. 27. The bit position to which the output is added depends on L. Initially, all bits are loaded with 1. At every clock cycle, the register contents is shifted to the right, the rightmost bit is applied to the output and the result of the modulo 2 addition is fed into the leftmost bit. The output sequence is repeated after 2^L-1 steps. For practical purposes, the elements b_k of the MLBS can be computed from a recursion formula:

$$b_{n+L} = \sum_{l=0}^{L-1} \alpha_l b_{n+l} \ , \tag{72}$$

$b_0 \ldots b_{L-1} = 1$ and the α_l are 0 or 1 and tabulated [78].

MLBSs have a white power spectrum with a minimum noise amplification, $U = 1$, and they might appear to be ideal excitation sequences. There are, however, subtle problems, which lead to signal 'echoes' as shown in Fig. 28. The position of the 'echo' depends on the length of the MLBS.

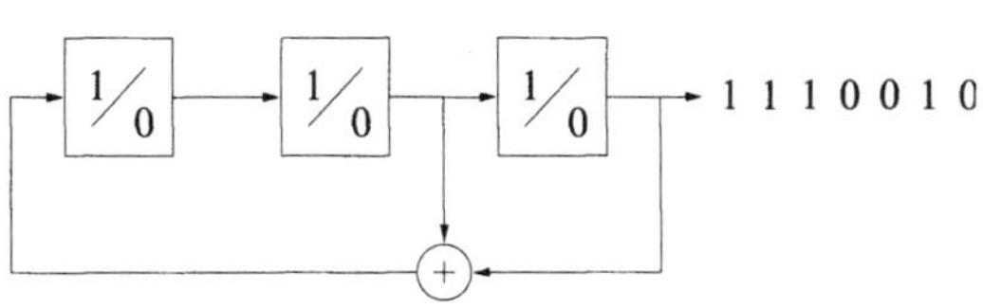

Fig. 27. 3-bit shift register to generate a maximum length binary sequence of length $2^3-1=7$. The summation is modulo 2

The reason for these localized perturbations stems from a slight delay and asymmetry in the switching characteristics of the Pockels cell [75]. As a consequence, *x'(t)* in Fig. 25 must be replaced by

$$x'(t) \propto x(t) + \varepsilon^{+} \frac{dx(t)^{+}}{dt} + \varepsilon^{-} \frac{dx(t)^{-}}{dt} \ . \tag{73}$$

The plus and minus sign symbolize the transitions from −1 to +1 and from +1 to −1, respectively.

5.6 Optimized Random Binary Sequences

Truly random binary sequences do not show the time-localized echoes of MLBSs shown in Fig. 28. The perturbation is spread instead as additional noise over the entire time window, which is less critical since there is no systematic deformation of *g(t)*. The noise amplification *U* of an arbitrary sequence is, however, unacceptably high. In the following, it will be shown how random sequences with low noise amplification and without the problems of MLBSs can be constructed.

Mapping to Ising model: An alternative view of the problem is obtained, when the pseudostochastic random binary sequence is interpreted as a spin-1/2 system. Every element $x[k]$ corresponds to a spin s_k that points either up or down (Fig. 29).

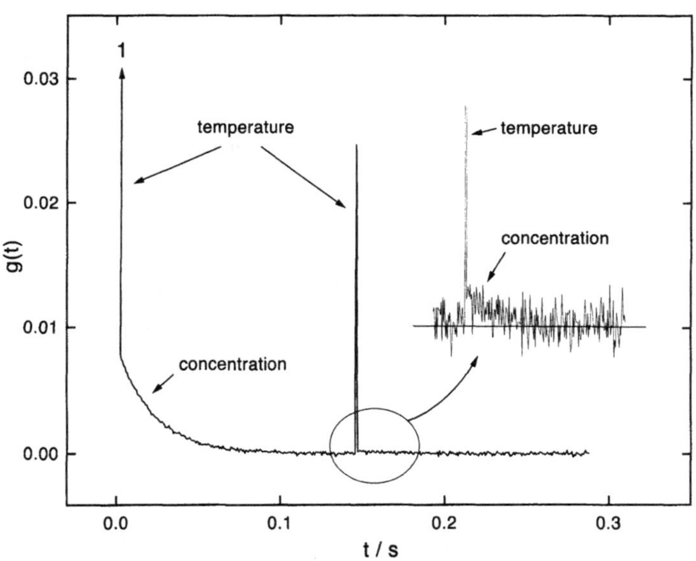

Fig. 28. Imperfections of the Pockels cell lead to 'echoes' in *g(t)* when measured with MLBSs

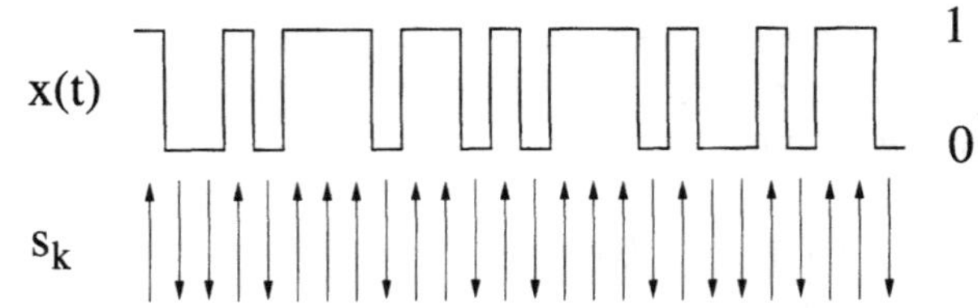

Fig. 29. The pseudostochastic binary sequence used to generate the excitation *x(t)* can be represented by a spin-1/2 system

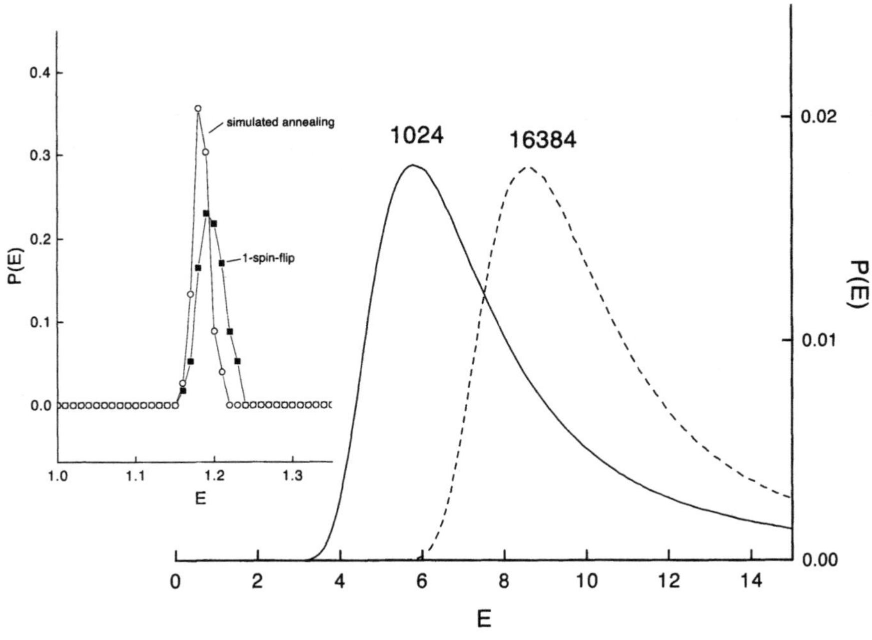

Fig. 30. Energy distribution of the spin system as obtained from randomly generated sequences of length 1024 and 16384 and after 1-spin-flip optimization with and without simulated annealing

The energy E of this spin system may be defined as the integral noise amplification U as given in Eq. (70), which can be expressed as a function of the pair products of all spins:

$$E = \sum_l \left(\sum_{k,m} a_{lk} a_{lm}^* s_k s_m \right)^{-1} \tag{74}$$

The coupling constants are the Fourier coefficients $a_{lk} = exp(-2\,\pi ikl/N)$. The task of finding a sequence with a low noise amplification is equivalent to the one of finding a minimum in the energy of all possible 2^N spin configurations.

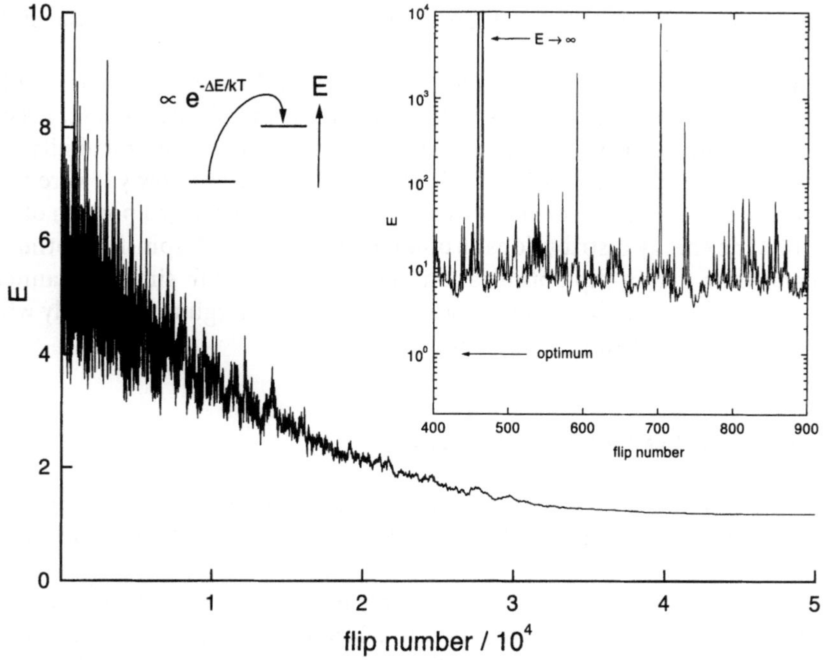

Fig. 31. Decrease of energy during simulated annealing. The temperature is exponentially lowered from $kT=1$ to 0. The insert shows a cross section through the energy landscape as a function of the number of stochastic 1-spin-flips

The energy landscape of this Ising spin system is very rugged, and there are even spin states where the energy diverges, as shown in the insert of Fig. 31. The distribution of energies $P(E)$ is very asymmetric with a long tail towards high E and a sharp cutoff at low E. The whole distribution is shifted to higher energies with increasing sequence length. Fig. 30 shows $P(E)$ for $N=1024$ and $N=16384$. The distribution for $N=1024$ has been obtained from 10^8 randomly generated sequences, each requiring a fast finite Fourier transform to compute E. The lowest value, which occurred only once in 10^8, is $E = 2.8$. This is still far above the best theoretical value of $E = 1$. Hence the chances of finding a suitable sequence by trial and error within a reasonable time are almost zero.

Optimization: Spin configurations with energies close to the global minimum, $E=1$, can be achieved by optimization of randomly generated seed sequences. Starting with the initial sequence, one spin after the other is flipped. If this results in a decrease of E, the spin is left in the flipped orientation, otherwise it is returned into its previous state. The procedure is repeated as long as the energy can be decreased. This simple and fast technique consistently yields $E \approx 1.2$, which is almost identical to the global minimum for all practical purposes. A slight improvement can be achieved by allowing also 2-spin-flips, which lead to lower energies if two spins are flipped simultaneously, but not necessarily for flips of only 1 of both spins.

Trapping in local minima of the rugged energy landscape can also be avoided by simulated annealing, which is well known from the physics of spin glasses [79]. Here, spin flips which increase E are allowed with a certain probability, given by a Boltzmann factor $exp(-\Delta E/kT)$. The temperature T is slowly lowered and the system frozen into lower energy states (Fig. 31). The enlarged portion of Fig. 30 shows the energy distributions as obtained from simple 1-spin-flip optimization and from 1-spin-flip optimization in combination with simulated annealing. The improvement by simulated annealing is only marginal, especially when compared to the increased computational effort.

5.7 Measurement with a White Power Spectrum

Figure 32 shows measurements of a bimodal PS with two very different molar masses of 0.895 kg/mol and 250 kg/mol at concentrations of 0.0166 and 0.0228, respectively [75].

The solvent was toluene. The same sample has been measured with PCS, TDFRS with a single long exposure pulse, and TDFRS with a pseudostochastic exposure sequence ($\Delta t = 140\ \mu s$, N=2047), which has been optimized by the 1-spin-flip optimization described in the previous section. The PCS experiment was originally measured with $\lambda = 647$ nm at an angle of 90° and has been shifted to

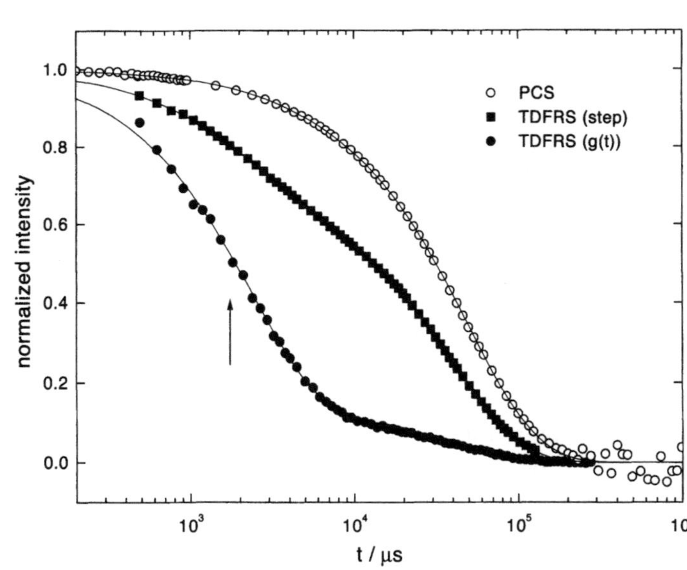

Fig. 32. Bimodal PS (0.895 and 250 kg/mol) as measured by PCS ($g_1(t)$), long exposure TDFRS (step) and TDFRS with pseudostochastic excitation ($g(t)$). The arrow marks the fast diffusion of the component with M=0.895 kg/mol. From Ref. [75]

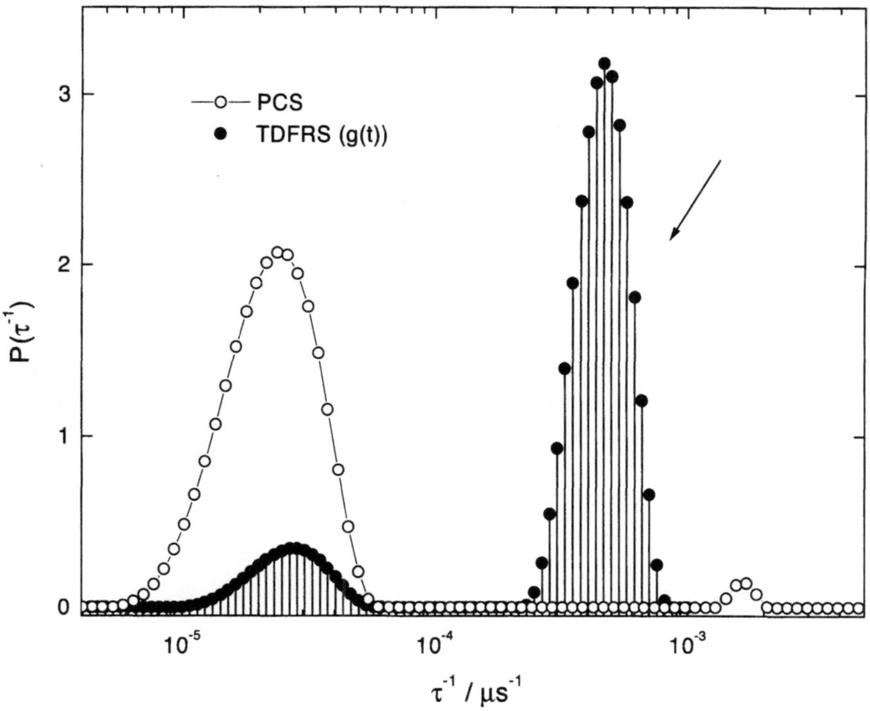

Fig. 33. Rate distribution of the PCS and the stochastic TDFRS (*g(t)*) measurement in Fig. 32. The arrow marks the fast diffusion of the component with *M=0.895* kg/mol. From Ref. [75]

the q-value of the TDFRS measurement according to $\tau \propto q^{-2}$. The total time needed for the measurements was the same in all cases. Even from a first visual inspection, the resolution of the two modes in the TDFRS measurements is obvious, whereas the PCS field autocorrelation function is completely dominated by the slow time constant of the high molar mass component.

Fig. 33 shows in a more quantitative analysis the rate distributions as obtained by a CONTIN analysis of the PCS and the stochastic TDFRS measurement.

The ratio of 6.8 for the two peak areas from stochastic TDFRS is close to the value of 5.9 as expected from the concentration ratio and the refractive index increments of the two PS, which depends on molar mass due to end-group effects. The thermal diffusion coefficient $D_T = 1.12 \times 10^{-7}\,\mathrm{cm}^2\,(sK)^{-1}$ is in excellent agreement with the value found previously in our laboratory [36].

The PCS rate distribution shows only the large peak of the slow mode, corresponding to the high molar mass component. The tiny peak at τ^{-1} appears at too high rates and is probably an artifact.

5.8 Colored Spectra and Tailored Excitations

Frequently, a flat power spectrum is not the best choice for a given problem, and it might be of advantage to focus the excitation energy into certain frequency ranges of interest. One example is a system with fast and slow relaxation processes of comparable amplitude and a large time scale separation of several decades. In such cases, a white power spectrum with a constant power density is inferior to a power spectrum, where the discrete frequencies are evenly spaced on a logarithmic instead of a linear frequency scale.

Mathematically, the problem is easily solved. Remember that the minimization of Eq. (70) is the criterion for a white power spectrum. If $C(\omega)$, or $C[k]$ in the discrete case, is the aimed-at colored power spectrum, then the pseudostochastic binary sequence whose power spectrum is the best approximation to $C[k]$ must fulfil $\sum_k C[k]\,|X[k]|^{-2} = minimum$. In the language of the the spin-1/2-system, the new energy function is

$$E = \sum_l C[l] \left(\sum_{k,m} a_{kl} a^*_{ml} s_k s_m \right)^{-1} . \tag{75}$$

The rest of the optimization is identical to the procedure described for the white power spectrum. The total excitation power in a (−1,1) binary sequence is, according to Parseval's theorem [Eq. (66)] always constant.

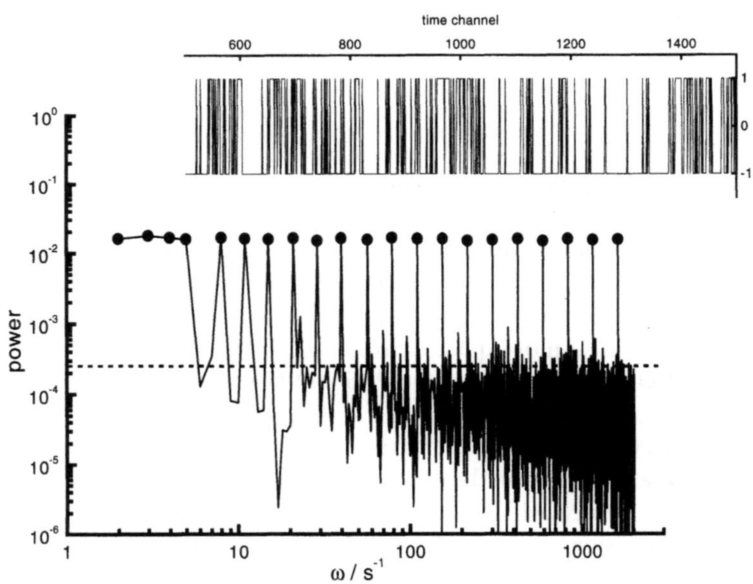

Fig. 34. Pseudostochastic binary excitation (upper part) with power concentrated at a small number of logarithmically spaced frequencies. The dashed line is the power density if the sequence had been optimized with a white power spectrum

Fig. 34 shows the power spectrum of an excitation sequence, which has been optimized with the energy function of Eq. (75) so that almost all excitation energy is found at a relatively small number of logarithmically spaced frequencies. To the eye, the sequence still looks random.

A measurement with this sequence is equivalent to the frequency domain experiment in Fig. 24, however with the multiplex advantage, i.e. all frequencies are measured simultaneously, not sequentially. The energy left at the intermediate frequencies can simply be discarded. The energies are still finite, since the aimed-at power spectrum can only be approximated during the minimization of *E*, as in case of the white power spectrum.

5.9
Digital Filtering

The concentration of excitation energy in certain frequency ranges does not change the measured linear response function *g(t)*. It only reduces noise at the frequencies of interest, at the expense of increased noise at the other frequencies. Not much is gained with respect to the noise amplitude in the time domain representation of *g(t)*. To reduce the increased noise at these unwanted frequencies, digital filtering is required. Such filtering also reduces the amplitude of signal components at those frequencies.

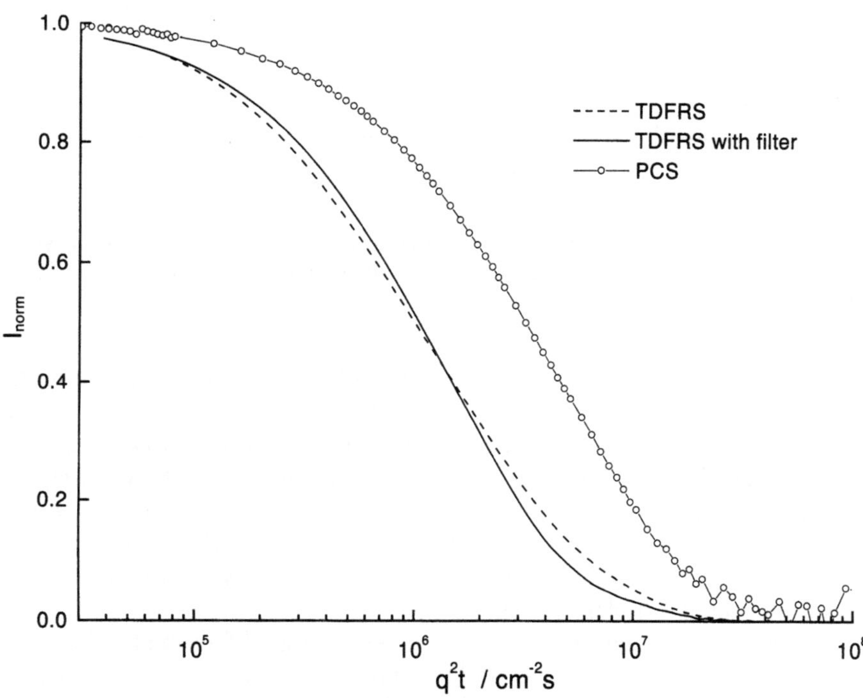

Fig. 35. Normalized decay curves for PCS and TDFRS with and without digital filtering

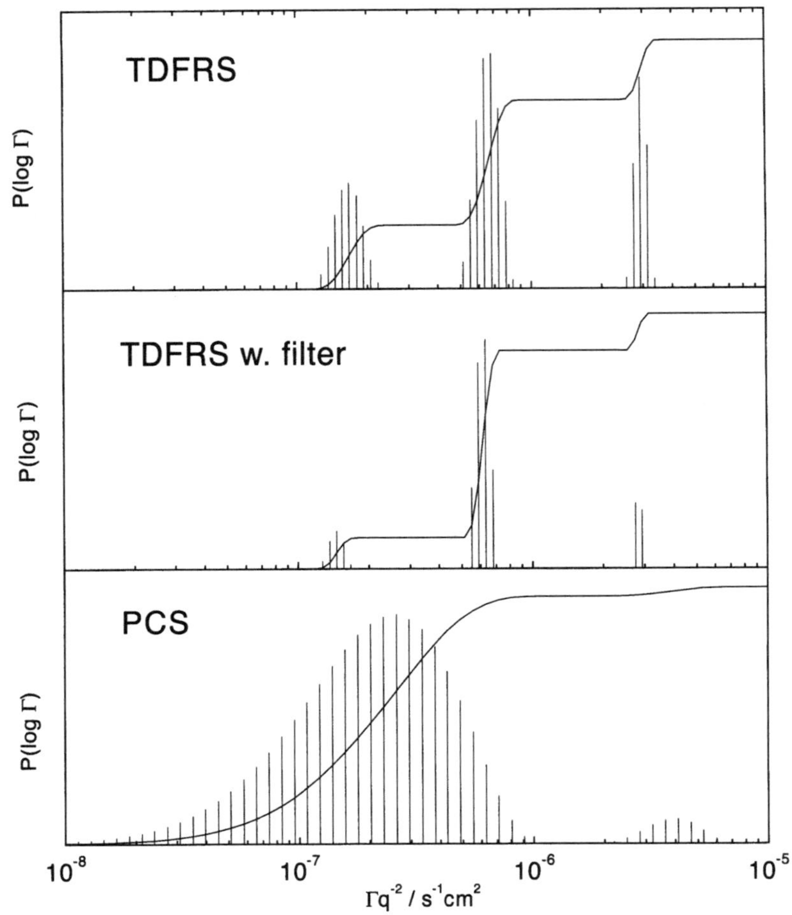

Fig. 36. Differential and integral rate distributions as obtained from a CONTIN analysis of the measurements in Fig. 35. The central peak is enhanced in the filtered stochastic TDFRS experiment

The result is very similar to the effect of the different statistical weights for the different molar masses when comparing TDFRS and PCS. The PCS correlation function $g_1(t)$ is dominated by the high molar masses. The low molar masses are suppressed, which means that signal intensity is shifted towards lower frequencies. The noise level is almost constant over the entire frequency range. Contrary to PCS, however, colored spectra together with digital filtering allow for an active selection of the frequencies of interest, and sensitivity can be tuned to e.g. low or intermediate molar masses.

Figures 35 to 37 show an experiment in which digital filtering has been utilized to enhance a certain species and to suppress others. The sample is a trimodal mixture of PS (5, 50, 2000 kg/mol) in toluene. It has been measured by PCS and stochastic TDFRS, with and without digital filtering.

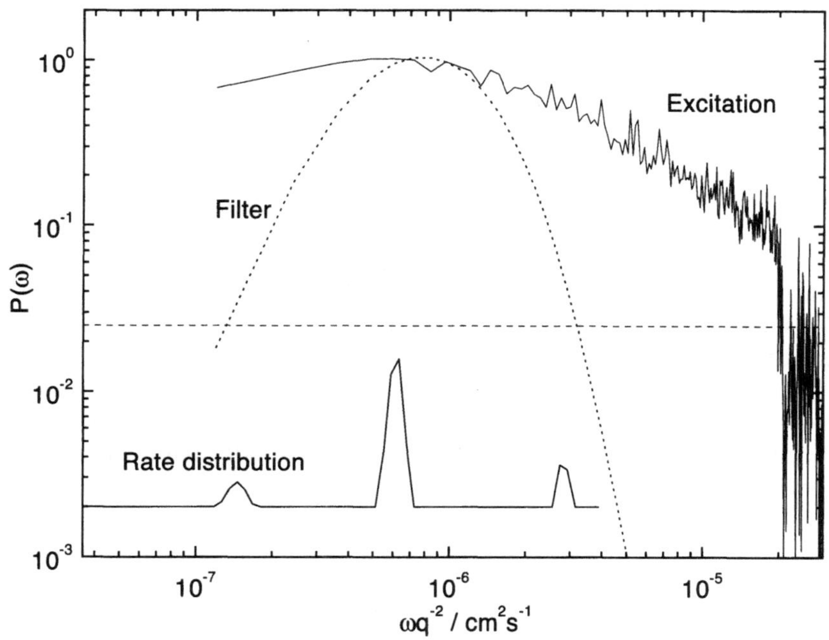

Fig. 37. Power spectrum of the excitation, filter function, and rate distribution for the filtered experiment in Figs. 35 and 36

Again the PCS signal decays more slowly than the TDFRS signal because of the dominance of high M in PCS (Fig. 35). This is also apparent from the rate distributions in Fig. 36. The PCS signal could not be resolved into three peaks. In the TDFRS signal, all three peaks are resolved and show approximately equal intensity.

First, an excitation sequence is designed by 1-spin-flip optimization of the energy, as defined in Eq. (75), whose power spectrum peaks around the rate of the middle process, corresponding to M=50 kg/mol, and falls off at lower and higher frequencies. Next, a digital filter F is constructed, which suppresses frequencies above and below the interesting frequency band. Figure 37 shows a sketch of the procedure with the excitation, the filter function, and the rate distribution.

The filtered response function is calculated according to

$$F\frac{YX^*}{|X|^2} = FH + FE\frac{X^*}{|X|^2} \quad . \tag{76}$$

Compared to the unfiltered response function in Fig. 35 the difference is obvious. The corresponding rate distribution in Fig. 36 shows a clear enhancement of the central peak and a strong suppression of the low- and high-M contribution. Finally, it should be noted that the same rate distribution could, in principle, be obtained without digital filtering, but much of the excitation energy would be sacrificed, and much of the good signal-to-noise ratio would be lost.

The question of how to construct the best excitation sequence for a given filter function is best answered when looking at the signal-to-noise ratio, which is in analogy to Eq. (71)

$$SNR = \frac{\langle |F|^2 |H|^2 \rangle}{\langle |F|^2 |E|^2 |X|^{-2} \rangle} . \tag{77}$$

It can easily be shown that for white noise with $|E[k]|^2 = \text{const.}$ and a random binary (–1,1)-sequence, whose total power is independent of its spectrum, Eq. (77) is maximized if the excitation spectrum is proportional to the power spectrum of the filter: $C[k] \propto |F[k]|^2$.

If the signal can be expressed as a sum over exponentials, as in the case of solutions of polydisperse polymers, high-, low-, and band-pass filters which are exponentials in the time domain influence the amplitudes, but not the diffusion time constants of the respective modes [80, 81].

6 Conclusion

We have outlined how TDFRS not only provides a useful tool for the study of the Ludwig-Soret effect in multicomponent liquids, but can also contribute valuable pieces of information towards solving the puzzles encountered in polymer analysis. Though TDFRS is conceptually simple, real experiments can be rather elaborate because of the relatively low diffraction efficiencies, which require repetitive exposures and a reliable homodyne/heterodyne signal separation. As an optical scattering technique it has much in common with PCS, and the diffusion coefficients obtained in the hydrodynamic limit ($q \to 0$) for monodisperse solutions are indeed identical.

The situation is different, however, in the case of solutions of polydisperse polymers. Here, TDFRS provides much more uniform statistical weights for the different molar masses. As a result, a better deconvolution into the rate and molar mass distribution can be achieved than in the case of PCS. The computation of the molar mass distribution and the molar mass averages does, however, require a knowledge of the scaling relation that relates D to M. Interestingly, the weight distribution of R_h can be obtained without knowledge of this scaling relation, which is not possible in the case of PCS.

While polydisperse model systems can nicely be resolved, the reconstruction of a broad and skewed molar mass distribution is only possible within certain limits. At this point, experimental techniques in which only a nonexponential time signal or some other integral quantity is measured and the underlying distribution is obtained from e.g. an inverse Laplace transform are inferior to fractionating techniques, like size exclusion chromatography or the field-flow fractionation techniques. The latter suffer, however, from other problems, like calibration or column-solute interaction.

An appealing aspect of TDFRS is the possibility to prepare some initial non-equilibrium state, whose time evolution is later observed. Especially through the application of modern signal-processing techniques, additional insight can be gained. Particularly easy to realize are noise-like pseudostochastic binary sequences. They provide a broad power spectrum, and the measurement may be viewed as a frequency-multiplexed lock-in technique. The most simple suitable sequences are characterized by a white power spectrum. After cross-correlation of the signal with the excitation, the linear response function is obtained, which eliminates all molar mass dependence of the individual signal contributions. Since arbitrary binary excitation sequences can be employed, it is also possible to tailor the corresponding excitation spectra according to specific experimental needs. Together with suitable digital filters, it is possible to selectively enhance or suppress certain frequencies and to amplify, for example, certain components in a multimodal sample.

References

1. Ludwig C (1856) Sitzber Akad Wiss Wien Math-naturw. Kl 20:539
2. Soret C (1879) Arch Geneve 3:48
3. Debye P, Bueche A M (1948) In: Robinson H A (ed) High-Polymer Physics Chemical Publishing, Brooklyn, p 497
4. Langhammer G H, Pfennig H, Quitzsch (1958) K Z Elektrochem 62:458
5. Ecenarro O, Madariaga J A, Navarro J, Santamaria C M, Carrion J A, Saviron J M (1990) J Phys Condens Matter 2:2289
6. Ecenarro O, Madariaga J A, Navarro J, Santamaria C M, Carrion J A, Saviron J M (1994) Macromolecules 27:4968
7. Tyrrell H J V (1961) Diffusion and Heat ow in Liquids Butterworth London
8. Emery A H, Drickamer H G J (1955) Chem Phys 23:2252
9. Hoffman J D, Zimm B H (1955) J Polymer Sci 15:405
10. Whitmore F C (1960) J Appl Phys 31:1858
11. Meyerhoff G, Nachtigall K (1962) J Polymer Sci 57:227
12. Nachtigall K, Meyerhoff G (1959) Makromol Chemie 33:85
13. Giglio M, Vendramini A (1977) Phys Rev Lett 38:26
14. Giglio M, Vendramini A (1975) Phys Rev Lett 34:561
15. Kolodner P, Williams H, Moe C (1988) J Chem Phys 88:6512
16. Zhang K J, Briggs M E, Gammon R W, Sengers J V (1996) J Chem Phys 104:6881
17. Giddings J C, Caldwell K D, Myers M N (1976) Macromolecules 9:106
18. Giddings J C, Hovingh M E, Thompson G H (1970) J Phys Chem 74:4291
19. Giddings J C, Caldwell K D, Myers M N (1976) Macromolecules 9:106
20. Brimhall S L, Myers M N, Caldwell K D, Giddings J C (1985) J Polym Sci Polym Phys 23:2443
21. Schimpf M E, Giddings J C (1989) J Polym Sci Polym Phys B27:1317
22. Venema E, de Leeuw P, Kraak J C, Poppe H, Tijssen R (1997) J. Chromatogr. A, 765:135
23. Schimpf M E, Giddings J C (1990) J Polym Sci Polym Phys B28:2673
24. M. Antonietti, A. Briel, and C. Tank. Acta Polymer., 46:254, (1995)
25. Kirkland J J, Rementer S W, Yau W W (1989) J Appl Polymer Sci 38:1383
26. Li W B, private communication
27. Thyagarajan K, Lallemand P (1978) Opt Commun 26:54
28. Köhler W (1993) J Chem Phys 98:660

29. Wesson J A, Noh I, Kitano T, Yu H (1984) Macromolecules, 17:782
30. Sillescu H (1988) Makromol Chem Makromol Symp 18:135
31. Hervet H, Leger L, Rondelez F (1979) Phys Rev Lett 42:1681
32. Rondelez F (1980) In: Degiorgio V, Corti M, Giglio M editors Light scattering in liquids and macromolecular solutions Plenum PubCorp New York p 243
33. Pohl D W, Schwarz S E, Irniger V (1973) Phys Rev Lett 31:32
34. Köhler W, Rossmanith P (1995) J Phys Chem 99:5838
35. Köhler W, Rossmanith P (1995) Int J Polym Analysis and Characterization 1:49
36. Köhler W, Rosenauer C, Rossmanith P (1995) Int J Thermophys 16:11
37. Rosenauer, Köhler W (1996) Macromolecules 29:3203
38. Johnson Jr C S (1985) J Opt Soc B 2:317
39. Eichler H J, Guenther P, Pohl D W(1986) Laser-induced dynamic gratings Springer Berlin Heidelberg New York
40. Späth H (1973) Algorithmen für multivariable Ausgleichsmodelle Oldenburg München
41. Holleman A F, Wiberg E (1971) Lehrbuch der anorganischen Chemie de Gruyter Berlin
42. Spill R, unpublished
43. Becker A, Köhler W, Müller B (1995) Ber Bunsenges Phys Chem 99:600
44. Allain C, Lallemand P (1977) C R Acad Sc Paris Ser B 285:187
45. Bloisi F, Vicari L, Cavaliere P, Martellucci S, Quartieri J, Mormile P, Pierattini G (1987) Appl Phys B44:103
46. Pohl D W (1980) Phys Rev 77A:53
47. Nagasaka Y, Hatakeyama T, Okuda M, Nagashima A (1988) Rev Sci Instrum 59(7):1156
48. Kogelnik H (1969) Bell System Tech J48:2909
49. Köhler W, Müller B (1995) J Chem Phys 103:4367
50. Li W B, Segre P N, Gammon R W, Sengers J V (1994) Physica A 204:399
51. Bou-Ali M M, Ecenarro O, Madariaga J A, Santamaria C M, Valencia J J (1998) J Phys Condens Matter 10:3321
52. Köhler W, Müller-Plathe F, Reith D Unpublished
53. Provencher S W (1982)Computer Phys Commun 27:213
54. Provencher S W (1982)Computer Phys Commun 27:229
55. Koppel D E (1972) J Chem Phys 57:4814
56. Chu B (1991) Laser Light Scattering Academic Press New York
57. Provencher S W (1979) Makromol Chem 180:201
58. Honerkamp J, Maier D, Weese J (1993) J Chem Phys 98:865
59. Schnablegger H, Glatter O (1991) Appl Opt 30:4889
60. Glatter O, Sieberer J, Schnablegger H(1991) Part Part Syst Charact 8:274
61. Stock R S, Ray W H (1985) J Polym Sci Polym Phys 23:1393
62. King T A, Treadaway M F (1977) J Chem Soc Faraday Trans 73:1616
63. Stölken S, Bartsch E, Sillescu H, Lindner P (1995) Prog Coll Polym Sci 98:155
64. Köhler W, Schäfer R, Bartsch E, Stoelken S (1997) Progr Colloid Polym Sci 104:132
65. Bartsch E, Frenz V, Sillescu H (1994) J Noncryst Solids 172:88
66. Stölken S, Bartsch E, Sillescu H, Lindner P (1995) Prog Coll Polym Sci 98:155
67. Renth F, Bartsch E, Kasper A, Kirsch S, Stölken S, Sillescu H, Köhler W, Schäfer R (1996) Progr Colloid Polym Sci 100:127
68. Medina-Noyola M (1988) Phys Rev Lett 60:2705
69. Kops-Werkhoven M M, Fijnaut H M (1981) J Chem Phys 74:1618
70. van Megen W, Ottewill R H, Owens S M (1985) J Chem Phys 82:508
71. M M Kops-Werkhoven and H M Fijnaut (1981) J. Chem. Phys., 77:2242
72. Beenakker C W J, Mazur P (1984) Physica A 126:349
73. Bloisi F Opt (1988) Commun 68:87
74. Schäfer R (1997) PhD thesis, Mainz
75. Schäfer R, Becker A, Köhler W (1998) Int J Thermophys 20:1 (1999)
76. Stearns S D (1987) Digitale Verarbeitung analoger Signale Oldenburg München

77. Ziessow D (1973) On-line Rechner in der Chemie de Gruyter Berlin
78. Ziessow D, Blümich B (1974) Ber Bunsenges Phys Chem 78:1168
79. Kirkpatrick S, Gelatt C D, Vecchi M P (1983) Science, 220:671
80. Stark H (1987) Image recovery theory and application Academic Press
81. S Stearns (1987) Digitale Verarbeitung analoger Signale Oldenburg München

Received: February 1999

Imaging of Polymers Using Scanning Force Microscopy: From Superstructures to Individual Molecules

Sergei S. Sheiko

Organische Chemie III / Makromolekulare Chemie,
Universität Ulm, D-89069 Ulm, Germany
e-mail: sergei.sheiko@chemie.uni-ulm.de

This review article describes progress made in scanning force microscopy of polymers during the last 5 years including fundamental principles of SFM and recent developments in instrumentation relevant to polymer systems. It focuses on the analytical capabilities of SFM techniques in areas of research where they give the most unique and valuable information not accessible by other methods. These include (i) quantitative characterisation of material properties and structure manipulation on the nanometer scale, and (ii) visualisation and probing of single macromolecules.

Keywords: scanning force microscopy, polymers, single macromolecules, thin films

List of Symbols and Abbreviations

AFM	atomic force microscopy
CFM	chemical force microscopy
DMT	Derjaguin-Muller- Toporov theory
FMM	force modulation mode
FIB	focussed ion beam
EM	electron microscopy
HOPG	highly oriented pyrolitic graphite
IC	intermittent contact
JKR	Johnson-Kendall-Roberts theory
LFM	lateral force microscope
μCP	microcontact printing
NMR	nuclear magnetic resonance
PCT	patch clamp technique
PFM	pulse force mode
SAM	self-assembled monolayer
SEM	scanning electron microscopy
SFA	surface force apparatus
SFM	scanning force microscopy
SNOM	scanning near field optical microscopy
SPFM	scanning polarisation force microscopy

SPM	scanning probe microscopy
STM	scanning tunnelling microscopy
TEM	transmission electron microscopy
UHV	ultra high vacuum
3D	three dimensional
a	contact radius
A	Hamaker constant, amplitude
b	width of the cantilever
D	distance between two surfaces
C	interaction constant
E	Young's modulus
f	interatomic force
F	force
h	thickness of the cantilever
g	gain factor of the optical lever
I	moment of inertia
k	harmonic spring constant
k_B	Boltzman constant
K	elastic modulus
m	mass of the cantilever
P	dissipation power
Q	quality factor
R	tip radius
r	interatomic distance, radius
r_{sp}	set-point ratio
S	spreading coefficient
V	volume
w	pair potential
w_a	work of adhesion
w_c	cohesion energy
W	interaction free energy
X	displacement of the tip parallel to the surface plane
Z	displacement of the tip perpendicular to the surface plane
Z_t	deflection of the cantilever (tip) from the equilibrium position
Z_c	position of the cantilever base relative to the ample surface
A	contact area
γ	surface energy
γ_{12}	interfacial energy
Π	surface pressure
φ	phase shift between the piezo drive and the cantilever response
ε	static dielectric constant
ω	frequency
ω_0	resonance frequency of the cantilever
ν	Poisson's ratio
δ	indentation depth of the sample

β damping coefficient of the sample
Δ sensitivity of the detection system

1 Introduction

Polymer surfaces and thin polymer films are distinctive fields of academic research dealing with a large body of technological applications. Polymers are widely used as main components in many industrial products in addition to being additives which stabilise colloidal suspensions [1–3], promote lubrication and adhesion [4, 5], and impart water and oil repellence [6, 7]. These applications of polymers have changed our life significantly during the last 20 years. More recently, polymers became building blocks in functional structures on ever decreasing length scales [8–18]. Nanometer sized templates, molecular devices, particulate films, and patterned surfaces are of increasing importance for advanced developments in microelectronics, catalysis, sensor technology, microbiology, and biomineralization [19–23].

Complementary to microfabrication processes, molecular concepts in nanotechnology rely on interfacial organisation and self-assembly of molecules yielding well-defined structures in thin films over large areas. Rigorous control of the molecular organisation on the nanoscale has set new demands on chemistry and characterisation methods including microscopy. Modern concepts in polymer chemistry are directed towards specific control of the macromolecular structure including degree of polymerisation, chain architecture, end functionalities and composition [24–27]. Examples are block copolymers, which yield regular and manifold microdomain structures ranging from 10–100 nm [28–33]; hyperbranched polymers, which demonstrate molecular segregation and surface interaction of individual molecules on the scale of 1–10 nm [34–38]; and finally, biopolymers with well defined tertiary structures [39–41].

Progress in nanotechnology also depends critically on new developments in microscopy [42–45]. Compared to other investigation methods that help to explore the relation between the molecular structure and macroscopic properties, microscopy gives the most direct information. Particularly, in the case of disordered or aperiodic structures, visualisation of the structure is often more useful than indirect measurement and interpretation of its scattering properties. In practice, the utilisation and value of microscopes depends on their spatial resolution, the contrast and the imaging conditions.

Three major advancements in resolution have occurred since Hookes's discovery of the optical microscope in 1665 [46]. In 1873, Ernst Abbe established fundamental criteria for the resolution limit in optical microscopy [47], which did not exceed the range of a couple of 100 nanometers even after the introduction of the confocal optical microscope [43, 48]. The invention of the transmission electron microscope by Ernst Ruska in 1933 extended the resolution of microscopes to the nanometer scale [49]. Finally, scanning tunnelling microscopy introduced, by Binnig and Rohrer in 1981, made a breakthrough when atomic resolution images were reported for the first time [50]. STM became the first member of a new class of scanning probe microscopes (SPM), which are based on a different principle than the optical and electron microscopy. Instead of lenses and electromagnetic waves, SPMs use a sharp probe which scans across the sample to sense different types of interactions in the near-field of the surface. For example, in STM, due to the strong exponential dependence of the tunnelling current on the distance between the probe and the surface (the current decreases by a factor of ten when the distance is increased by just 1 Å), a resolution of ca. 0.01 nm can be achieved.

There are other important advantages to be considered. In addition to the lateral resolution, SPMs are capable of three dimensional imaging of the surface morphology and can provide a topographic map of the sample. Owing to the new operation principle, SPMs show few limitations regarding imaging conditions. Sample preparation became easier compared to the electron microscope. The latter often requires rather sophisticated sample treatment such as etching and metal sputtering, while SPMs can visualise the native structure of the surface. For many years, biologists have struggled to combine the high-resolution advantages of the electron microscope with the in-water operating capabilities of the optical microscope. Currently, most SPMs can operate in different environment including vacuum, air and liquid. This was a considerable advance which allowed in-situ observation of biological systems in their natural medium. Finally, SPM set up requires a much lower capital investment, as well as the maintenance costs compared to electron microscopy.

Depending on the probe type, one can distinguish between Scanning Tunnelling Microscopy (STM), Scanning Force Microscopy (SFM), Scanning Near Field Optical Microscopy (SNOM), and other S(*X*)M techniques, where *X* refers to physical properties such as magnetic field, electrostatic charge, and temperature [51–57]. STM, SFM and SNOM are the most widely used methods so far. Each of these microscopies, however, shows certain limitations for polymer characterisation. STM is based on measuring the tunnelling current (picoamps to nanoamps) that flows through an electronic gap between a conductive tip and a conductive sample. For non-conductive materials, however, STM is of limited use. Eventually, the microstructure of thin organic films was investigated by STM down to atomic resolution [58–61]. In most cases, STM is not applicable for polymer films, whose thickness is determined by the chain conformation whose dimensions are typically >>1 nm exceeding the tunnelling gap by far.

SNOM combines the optical contrast with a high lateral resolution of SPMs [55, 56]. Scanning a surface with a sharp optical fibre tip within the range of the optical near field makes it possible to overcome the optical diffraction limit that restricts the resolution of conventional optical microscopy. Moreover, the SNOM probe operates at a finite distance from the surface, so that damage and distortion of delicate samples can be eliminated. The drawback of SNOM compared to other SPM methods is its relatively low resolution – around tens of nanometers [62, 63].

The first of the scanning probe microscopies which overcame the limitation of STM in imaging of non conductive materials, however, was atomic force microscopy (AFM) introduced by Binnig et al. [64]. It soon became a versatile method for characterisation of polymer surfaces and thin films renamed as scanning force microscopy. In addition to the favourable imaging conditions and the high resolution, SFM offers a variety of new contrast mechanisms which can distinguish surface areas differing in chemical and physical structure. Thus, the SFM measurements extend the microscopic capabilities far beyond morphological studies. Now mechanical properties, such as visco-elasticity, friction, and adhesion, as well as long range electrostatic and steric forces can be characterised quantitatively at the scale of a few nanometers. Furthermore, measurement of molecular forces and manipulation of single macromolecules, something not possible with optical and electron microscopy, are becoming feasible.

The main objective of this review article is to demonstrate the analytical capability of SFM techniques in areas of research where SFM gives the most unique and valuable information not accessible by other methods. Complementary to other reviews [65–69], we will focus on recent developments in SFM instrumentation which are particularly useful for polymer systems (Sect. 2); on quantitative characterisation of material properties and structure manipulation on the nanometer scale (Sect. 3); and on visualisation and probing of single macromolecules (Sect. 4). The interested reader can find the theoretical background as well as instrumentation of SFM in text books [52–54, 70–72].

2 Fundamentals of Scanning Force Microscopy

The main purpose of the invention of the first SFM was to extend the imaging capabilities of STM to non-conductive samples [64]. This was enabled by sensing electromagnetic forces that act in the near field of the surface (Fig. 1a). In prac-

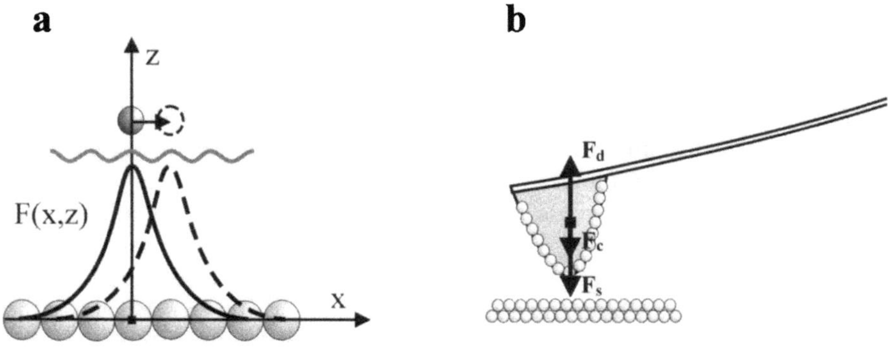

Fig. 1. a Variation of the interaction force between a flat surface and an isolated atom in the near field of the surface. The dot line corresponds to the lateral displacement of the atom by one interatomic distance. The grey line indicates the force profile "sensed" by the atom as it moves parallel to the surface plane. **b** Scheme of a SFM probe: a sharp tip mounted on a cantilever. The interaction force $F_i=F_s+F_d$ is a sum of many interatomic interactions, where F_s is the surface force and force F_d results from the sample deformation. The interaction force is balanced by force F_c due to the cantilever bending

tice, it was realised by measuring the mechanical interaction between a sharp tip and the sample to be investigated (Fig. 1b).

Conceptually, predecessors of the scanning force microscope are the surface force apparatus (SFA) [73, 74] and the stylus profilometer [75, 76]. The SFA enables measurement of normal and friction forces between molecularly smooth surfaces as small as 1 nN as a function of distance with a resolution of 0.1 nm. In addition to the local force measurement, the profilometer provides a topographic map of the surface by scanning the surface with a sharp probe. However, the profilometer is not suitable for structure characterisation because of the large radius of the tip (about 1 μm) and the low sensitivity of the force sensor (in the range of 10^{-2} to 10^{-5} N).

SFM combines the high force sensitivity of the surface force apparatus with the scanning principle of the profilometer. It enables measurement of weak forces as low as 1 pN with a spatial resolution down to 1 Å.

2.1 Intermolecular and Surface Forces Relevant to SFM

A force microscope actually measures the forces between two macroscopic bodies. The finite size and the macroscopic surface of the tip and the surface spot lead to a number of fundamental consequences in their interaction (Fig. 2). First, the net force is stronger than the intermolecular forces and it acts at much larger distances. Even in the 10–100 nm range, the interaction energy, which is proportional to the size of the tip, can exceed k_BT. Secondly, the force between a spherical tip and a flat surface decays with the separation as $F{\sim}D^{-2}$ (Fig. 2b) compared to $f{\sim}r^{-7}$ for the attraction between two atoms (Fig. 2a). In combination with the finite tip size, the low force gradient increases the effective interaction area and limits the resolution (see Sect. 2.3.3). Third, the surrounding me-

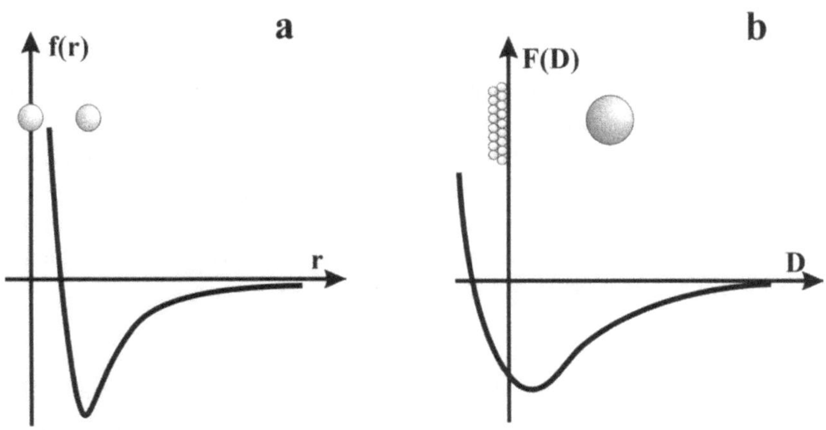

Fig. 2. Interaction forces acting in vacuum between **a** two atoms ($f{\sim}r^{-7}$) and **b** macroscopic particles (e.g., for surface-sphere interaction, $F{\sim}D^{-2}$). The tip position at D=0 corresponds to the tip-sample contact, while the range at D<0 corresponds to the sample indentation

dium can modify the interaction potential resulting in local minima. And finally, the deformation of the bodies upon contact increases the contact area and results in additional contribution to the net force.

In general, the forces which contribute to the net force exerted on the tip can be divided into three groups (Fig. 1b): (i) surface forces, F_s, (ii) forces due to the sample deformation, F_d, and (iii) the deflection force of the cantilever, F_c. It is important to note that all three forces can be of either sign. The van der Waals force is determined by the Hamaker constant, which depending on the dielectric properties of the contacting phases can be positive or negative. Also the deformation force can be repulsive or attractive, when the sample is pressed or stretched by the tip, respectively.

Let us first consider the **surface attraction.** An elementary constituent of the interaction between a flat, rigid surface and a sharp, rigid tip in vacuum is the pair potential between atoms or small molecules at the tip and the sample, respectively. The intermolecular forces are essentially electrostatic in origin and can be calculated from the spatial distribution of the electron clouds. At large distances the forces are attractive and are described by the van der Waals potential $w(r)=-C/r^6$, where C is the interaction constant determined by polarizability and dipole moments of the molecules. In order to relate the atomic interaction to the interaction between the macroscopic tip and macroscopic surface, one has to sum up all intermolecular potentials involved in the interaction. A simple integrating procedure can be done assuming additivity of the intermolecular forces [77]. For a sphere-surface potential, which is a good approximation for the interaction between the tip and sample, the attractive part of the interaction energy becomes

$$W(\mathrm{D})=-\mathrm{AR/D}, \tag{1}$$

where A is the Hamaker constant, R is the radius of spherical tip, and Z is the tip-surface distance. This gives the attractive force:

$$\mathrm{F_a}=\mathrm{d}W/\mathrm{dD}=\mathrm{AR/D^2} \tag{2}$$

For a typical value of the Hamaker constant in vacuum, $A=10^{-19}$ J, the attractive force emerging between a tip with an apex radius of 10 nm and a surface separated by 1 nm distance will be F=1 nN. This value sets an approximate scale of the forces which are sensed by the scanning force microscope.

When the tip and sample come in physical contact (Fig. 3a), i.e. D≅intermolecular distance, the surface interaction from Eq. (2) corresponds to the adhesion force which is related to the surface energy γ as

$$\mathrm{F_{adh}}=4\pi\mathrm{R}\gamma \tag{3}$$

This equation assumes that the tip and sample are of the same material and no deformation occurs. If the tip and the surface are made of different materials, the cohesion energy $w_c=2\gamma$ should be replaced by the work of adhesion. For a hydrocarbon polymer, where mainly dispersion forces are responsible for the tip-surface interaction, the work of adhesion can be estimated as $w_a=2(\gamma_t{}^d\gamma_s)^{1/2}$, where γ_t^d denotes the dispersion part of the tip surface-energy and γ_s is the surface energy of the sample [77]. For a Si tip and a polymer surface (γ_t^d=100 mJ/m^2 and γ_s=25 mJ/m^2, respectively) the adhesion force will be about F_{adh}=6 nN. The value is not small and can cause severe deformation of the tip as well as the sample (Sect. 2.3.2).

Under ambient conditions, the atmosphere contains water and organic contamination. Depending on the relative humidity, water can condense around the contact site and result in capillary forces (Fig. 3b). The meniscus curvature varies with the relative vapour pressure and the tip shape [77,78]. For a small spherical tip, the capillary contribution to the adhesion force can be calculated as

$$F_{cap}=4\pi R\gamma_l\cos\Theta, \tag{4}$$

where γ_l is the surface tension of the condensing vapour and Θ is the contact angle. For water with γ_l=73 mN/m^2, the capillary force will be F_c=9 nN.

Additional complications can arise when two bodies, i.e. the tip and the sample, interact in liquid (Fig. 3c). The interaction energy of two macroscopic phases across a dielectric medium can be calculated based on the Lifshitz continuum theory. In contact, when the distance between the phases corresponds to the nonretarded regime, the Hamaker constant in Eq. (2) is approximated by:

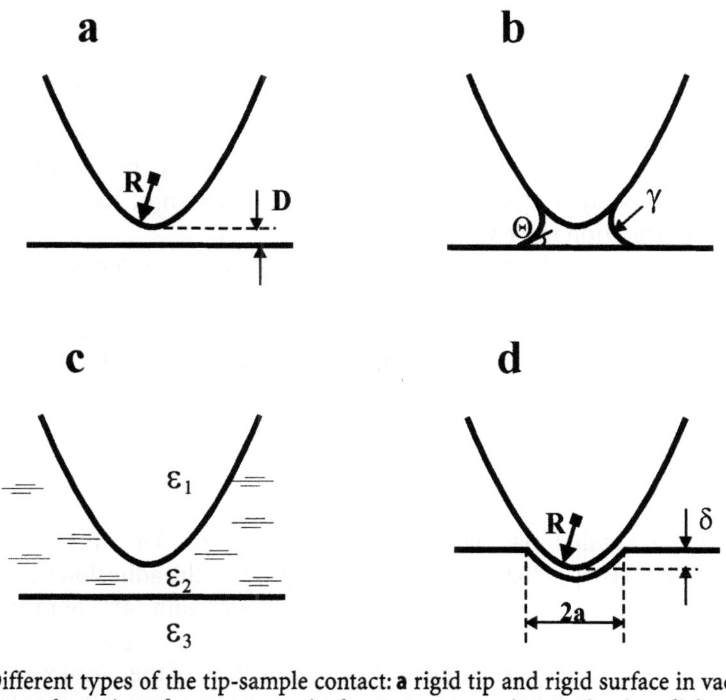

Fig. 3. Different types of the tip-sample contact: **a** rigid tip and rigid surface in vacuum, **b** capillary condensation of water vapour in the contact area, **c** interaction in a dielectric medium, **d** deformation of a soft sample induced by a rigid tip

$$A \approx \frac{3k_B T}{4} \frac{(\varepsilon_1 - \varepsilon_2)(\varepsilon_3 - \varepsilon_2)}{(\varepsilon_1 + \varepsilon_2)(\varepsilon_2 + \varepsilon_3)} \tag{5}$$

If the dielectric constant of the medium ε_3 is in between the dielectric constants of interacting phases 1 and 2, i.e. $\varepsilon_1<\varepsilon_3<\varepsilon_2$ or $\varepsilon_1>\varepsilon_3>\varepsilon_2$, the Hamaker constant is positive and favours attraction. If ε_3 is either smaller or larger than ε_1 and ε_2, the Hamaker constant becomes negative resulting in repulsion between the tip and sample. In addition, the surrounding medium can contain ions and dissolved molecules. This can change the interaction potential in a complicated fashion, depending on the molecular composition, pH and ionic strength of the medium [77].

At shorter distances the **repulsive forces** start to dominate. The repulsive interaction between two molecules can be described by the power-law potential $\sim 1/r^n$ ($n>9$) caused by overlapping of electron clouds resulting in a conflict with the Pauli exclusion principle. For a completely rigid tip and sample whose atoms interact as $1/r^{12}$, the repulsion would be described by $W\sim 1/D^7$. In practice, both the tip and the sample are deformable (Fig. 3d). The tip-sample attraction is balanced by mechanical stress which arises in the contact area. From the Hertz theory [77, 79], the relation between the deformation force F_d and the contact radius a is given by:

$$F_d=Ka^3/R \tag{6}$$

where K is the elastic modulus of the tip-sample contact $K = \frac{4}{3}\left[\frac{1-\nu_t^2}{E_t} + \frac{1-\nu_s^2}{E_s}\right]^{-1}$,

where E_t, ν_t and E_s, ν_s are the Young's moduli and Poissons's ratios of the tip and sample, respectively. For a typical contact radius a=5 nm, K=1 GPa and R=10 nm, a deformation force will be F_d=12.5 nN.

In these equations, the repulsion of the sample became interrelated with the adhesion force via the contact area. Several models have been developed to include the effect of the adhesion forces [80–83]. Johnson, Kendall, and Roberts derived the following expression for the contact radius and surface indentation:

$$a = \left(R/K\right)^{1/3} \left(F + 3\pi R w_a + \sqrt{6\pi R F w_a + (3\pi R w_a)^2}\right)^{1/3} \tag{7}$$

where F is the applied load, w_a is the work of adhesion. The important consequence of this theory is that even at zero load (F=0 N) the adhesion between the tip and sample leads to a finite contact area and a certain sample indentation. These are the two major factors which limits the resolution in SFM (Sect. 2.3.3).

The interaction forces are balanced by the deflection of the cantilever $F_c=k\Delta z$. The spring constant of the cantilever depends on the lever geometry and the material used to built the cantilever. For example, the spring constant of a massive

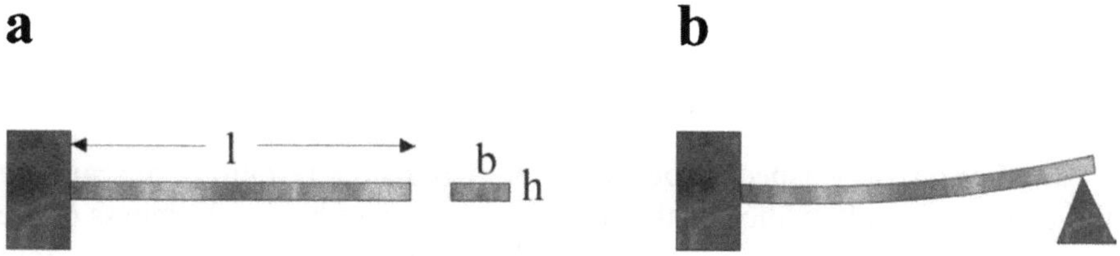

Fig. 4. Two configurations of massive levers which are the most relevant to SFM. **a** free cantilever of the length l, the width b and the thickness h; **b** cantilever supported at the end as the tip contacts the surface

cantilever beam with a rectangular cross section ($b \times h$) in a free (noncontact) state is $k=3EI/l^3$, where E is the Young's modulus and I is the momentum of inertia $I=bh^3/12$ (Fig. 4a). For a cantilever in repulsive contact with the sample (Fig. 4b), the spring constant becomes larger:

$$k=48EI/l^3 \tag{8}$$

For lateral force measurements, the constant of the cantilever torsion can be calculated from $k=GJ/l$, where G is the shear modulus and J is the polar moment of inertia:

$$J \approx b^3h^3 \Big/ 3.6(b^2+h^2)$$

So far a number of important conclusions can be summarised. The SFM signal (deflection of the cantilever) results from a combination of surface and deformation forces. The total surface force between a SFM tip and a polymer surface includes adhesion and capillary forces, $F_s=F_{adh}+F_{cap}$. For polymer surface, it can be estimated to $F_s \approx 15$ nN. Lower values might be expected for nonwettable and lower energy surfaces. The surface force is balanced by forces from the surface deformation, F_d, and from the deflection of the cantilever, F_c. In order to monitor a signal in contact SFM, the deformation forces must exceed the total surface forces, $F_c=F_d-F_s$. E.g., a net repulsive force of 0.4 nN will be monitored by 1 nm deflection of a Si cantilever with the following parameters E=50 GPa, l= 100 μm, b=30 μm, h=4 μm, k=0.4 N/m.

Note, that the surface and deformation forces are of the same order of magnitude. Therefore, surface forces should be as small as possible to minimise damaging and indentation of soft polymer samples. For example, sharp probes have a lower capillary attraction and adhesion forces, and therefore enable more gentle probing of a soft polymer than a blunt tip. A sharp tip can also be moved in and out of the contamination layer more readily than a blunt tip. This is particularly important for non-contact intermittent contact imaging described in Sect. 2.2.1.

Furthermore, the surface and deformation forces become interdependent via the contact area and respond nonlinearly to the displacement of the cantilever base. So far, most of the SFM images presented in literature depict a mixture of the morphology and material properties. It is a challenging experimental task to measure the forces in the net response of the cantilever separately and extract specific information about the surface structure. This will be discussed in Sect. 2.2.2.

Another puzzling aspect to be addressed regards the resolution in scanning force microscopy. In principle, the resolution is limited by the sharpness of the scanning probe. The best tips available so far have a probe apex radius of about 5 nm, which sets the lower limit in the resolution around a few nanometers. Nevertheless, images with an atomic resolution were readily achieved by many groups world-wide, though the contact diameter from Eq. (7) exceeds the intermolecular distance by far, e.g. $2a \cong 10$ nm. It became a routine operation to obtain atomic resolution images of polymers such as PE and PTFE [58, 84, 85]. However one has to discriminate clearly between true atomic resolution, that would also allow to visualise local defects, and the usually observed crystalline lattices (Sect. 2.3.3).

2.2 Instrumentation

Scanning of an SFM probe across the investigated surface is achieved by means of a piezoelectric scanner which moves the sample (or the tip) to generate a scan

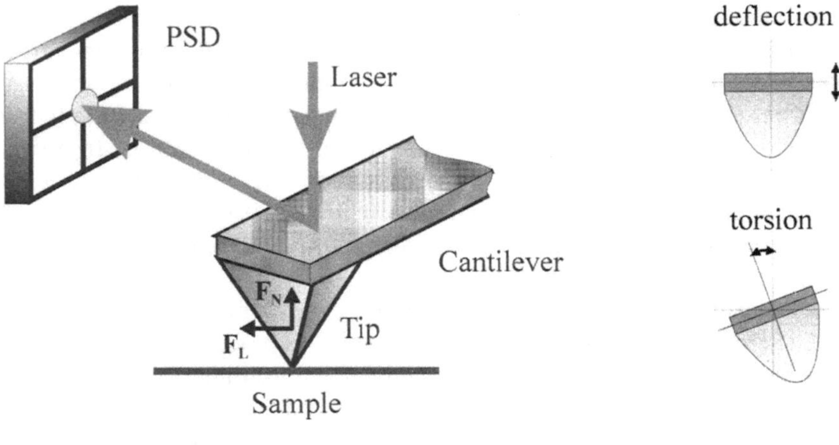

Fig. 5. a Schematic representation of the SFM set-up using the optical beam deflection method. **b** When the tip interacts with the sample surface, the cantilever exhibits deflection perpendicular to the surface as well as torsion parallel to the surface plane. The normal force F_N and the lateral force F_L corresponds to the force components which cause the deflection and torsion, respectively. Both responses are monitored simultaneously by the laser beam which is focused on the back side of the cantilever and reflected into a four-quadrant position sensitive detector (PSD)

in the X-Y plane. The scanner is usually fabricated from lead zirconate titanate, or PZT ceramic, which changes dimensions when a potential is applied to it, providing smooth displacement on the order of 1 Å per volt. Force measurements are made by recording the deflection of the cantilever end while its fixed end (cantilever base) is either moved towards or retracted from the sample. Currently, an optical lever system is the most common force sensor used in SFM. Cantilever deflections are measured by reflecting a laser beam at the free end of the cantilever (Fig. 5) [86, 87]. The change in the reflection angle is detected by a multiple segment photodiode known as a position sensitive detector (PSD). Other detection systems are discussed in specialised reviews [51–53].

2.2.1
Operating and Imaging Modes

Scanning force microscopes can be operated in many ways measuring different interactions between the probe tip and sample and using different types of detection schemes. Development of new modes is driven by the wish of the experimentalist (i) to discriminate the different contributions to the net force, such as adhesion, elasticity and viscosity, (ii) to enhance the resolution and contrast of the microscope, and (iii) to diminish distortion of the tip and sample during scanning.

Operating modes can be roughly classified as "contact", "non-contact", and "intermittent contact" modes depending on the sign of the net forces exerted on the tip. When the SFM is operating in the attractive region, it is called "non-contact". In this region, the cantilever is bent toward the sample, since it is being pulled by attractive forces. Operation in the repulsive region is called "contact" imaging. Here, the cantilever is bent away from the sample due to the sample deformation. If an oscillatory tip displacement is sufficiently large to pass through both regions, the probe experiences both attractive and repulsive forces. These modes are called "intermittent contact".

SFM's can be also classified according to static and dynamic operating modes. Under quasi-static conditions, the microscope measures the instantaneous response of the cantilever when it interacts with the sample. Dynamic SFM enables separation of the elastic and inelastic component in the cantilever deflection when the sample surface is exposed to a periodically varying stress field. The dynamic modes are useful for investigation of viscoelastic materials such as polymers and results in additional improvements in the signal-to-noise ratio.

SFM's can be also classified according to the nature of the medium introduced between the tip and sample, e.g. air (ambient SFM), ultra-high vacuum (UHV-SFM) and liquid (fluid SFM). In this section, we will discuss each of these aspects in more detail.

2.2.1.1
Contact SFM

In the contact mode, the probe senses forces F acting both perpendicular and parallel to the surface plane, i.e. *normal forces* and *lateral forces*, respectively (Fig. 5). Figure 6 shows a typical force-distance curve obtained when the tip is brought into close proximity with the sample surface. The normal force is proportional to the deflection of the tip(Z_t), while the distance is given by the position of the cantilever (Z_c) relative to the sample surface. Z_c=0 corresponds to the straight cantilever (Z_t=0) in contact.

The repulsive forces result in a vertical deflection of the cantilever away from the surface. For a hard sample, that does not deform, displacement of the cantilever base, ΔZ_c, will be equal to the tip deflection, ΔZ_t. If the spring constant of the cantilever, k_c, is known, one can determine the force exerted on the sample as $F=k_c\, Z_t$. With a typical SFM cantilever, which has a spring constant of 0.01 to 0.1 N/m, the 0.1 nm sensitivity of the optical lever sensor [87] allows registration of forces as low as 1 pN.

The surface structure can be imaged at constant height or at constant force (Fig. 7). In the "constant height" mode the vertical position of the cantilever base (Z_c) is kept fixed as the tip scans laterally across the sample and the cantilever deflection is measured to monitor the force variation dZ_t/dx. This imaging mode gives better results on relatively flat areas, where the dZ_t/dx signal allows one to identify tiny corrugations and atomic steps. By maintaining the normal force, i.e. deflection Z_t, at a constant level, the microscope measures variation in height (ΔZ_c), i.e. surface topography. This so-called "constant force" mode needs a

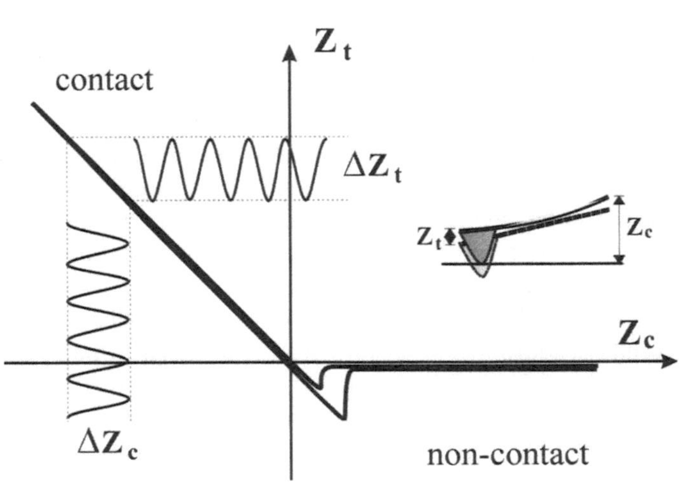

Fig. 6. A force-distance curve displays vertical bending of the cantilever (Z_t) versus displacement of the cantilever base relative to the sample surface (Z_c). Respectively, ΔZ_t and ΔZ_c shows the modulation of the cantilever response and the surface position during scanning of the surface by contact mode SFM

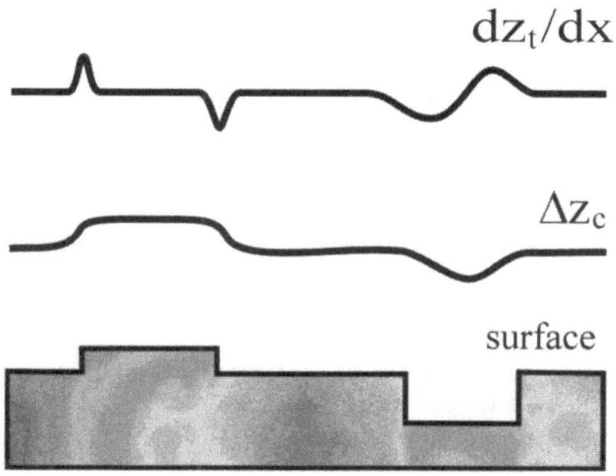

Fig. 7. Force variation (dz_t/dx) and topographic (Δz_c) profiles recorded by contact mode SFM maintaining a constant height (Z_c=constant) and a constant force (Z_t=constant), respectively. The actual surface structure is shown beneath the profiles

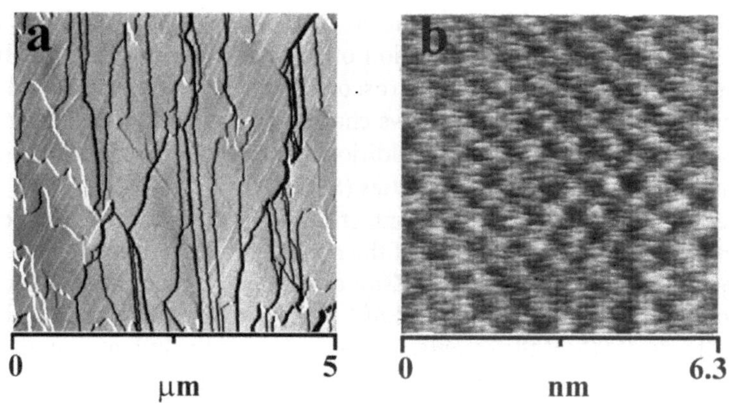

Fig. 8. Surface structure of a single crystal of perfluorododecyleicosane $F(CF_2)_{12}(CH_2)_{20}H$ measured by contact mode SFM on the **(a)** micrometer and **(b)** nanometer scale. The constant height image in **(a)** shows variation of the normal force, while the constant force image in **(b)** displays variation in height caused by densely packed CF_3 end groups protruding at the surface

highly efficient feedback loop and requires a low scan rate. At high speed, the topography signal is superimposed by cantilever deflection effects because of a relatively slow feedback response. For normal operating conditions, SFM contact imaging is a mixture of both modes.

One of the main advantages of contact SFM is the high spatial resolution. Variations in height as small as 0.01 nm can be easily detected in the constant height mode. Figure 8a shows as an example the image of the smectic structure of per-

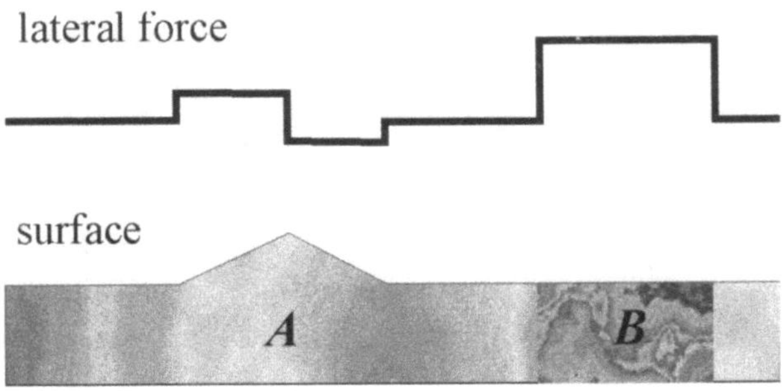

Fig. 9. Torsion of the cantilever, i.e. the lateral force signal F_L, upon scanning from left to right a corrugated surface **a** with a "sticky" region **b**

fluorododecyleicosane $F(CF_2)_{12}(CH_2)_{20}H$ where sharp layer steps cause strong variations of the force signal. Figure 8b shows a high resolution surface image with the regular arrangement of the CF_3 end groups.

The lateral force microscope (LFM) is a modification of the standard contact mode SFM [87–90]. In addition to the normal forces, the friction forces exerted on the probe are measured via torsion of the cantilever (Fig. 5). This mode is sometimes called "friction force microscopy". LFM can be used in combination with topographic imaging as it shows changes in material as well as enhanced contrast on sharp edges (Fig. 9). In addition to morphology, it provides information on the friction and wear properties (Sect. 3.4).

Other valuable extensions of contact-mode SFM probe micromechanical properties of the sample. Variation of the repulsive force or upward deflection of the cantilever are registered while a Z-modulation is applied either to the sample or the cantilever base (Fig. 6). The dZ_t/dZ_c signal contains information about the local stiffness $dF/d\delta$ of the sample, where $\delta = Z_t - Z_c$ is the indentation depth. Micromechanical applications will be discussed in Sect. 2.2.

2.2.1.2
Non-Contact SFM

A severe drawback of the contact mode is that the native structure of soft samples can be easily distorted due to adhesion and friction forces (Sect. 2.3.2). One way to overcome this problem, at least partially, is probing the surface in the non-contact regime. Operation of the non-contact mode is similar to the contact mode, except that the probe operates in the attractive force region, where the cantilever is bent towards the surface [91]. However, when the probe is not in contact with the surface, it is difficult to achieve stable imaging conditions. Eventually, the cantilever can be trapped by meniscus forces due to a liquid contamination layer between the surface and the probe. Because of instantaneous changes in the surface roughness, electrostatic charges at the surface, and the relative humidity in the surrounding atmosphere, the attractive forces can alter significantly. For example, if the probe breaks the meniscus as it crosses a high step on the surface, the feedback control will be lost and the tip will snap into contact with the sample surface.

Relatively stable imaging with a static non-contact mode can be performed by monitoring long range forces. For example, electrostatic forces have a much longer range compared to the van der Waals forces, and can be easily used for non-contact mode SFM [92]. The sensing of the electrical double layer enabled in-situ observation of hemicylindrical micelles of ionic surfactant molecules on graphite at a separation of ca. 40 nm [93].

To enhance the sensitivity to weak forces acting at large distances, a modulation technique has been developed. In this mode, the cantilever is vibrated at its resonant frequency with a constant amplitude as it is brought close to the sample surface. The amplitude is set to be smaller than the tip-sample separation, e.g. about 1 nm, so that the force gradient can be sensed at different distances from the surface without contact to it. When the cantilever senses a positive force gradient, the resonance frequency shifts to lower values. A feedback loop is used to adjust the distance by maintaining a constant frequency. This way, a three-dimensional image of the sample surface is generated. Force gradients as small as 10 mN/m and forces as small as 0.1 pN were detected at a tip-surface separation gap of 1 nm [91, 94, 95]. Recent developments in non-contact SFM has enhanced the achievable resolution significantly and made it possible to resolve monoatomic defects [96]. In these experiments, the extremely stable conditions over long time periods could be only achieved in ultra high vacuum at temperatures down to 10 K.

Although it is a prominent advantage of non-contact SFM, that images may be obtained of very soft samples, the techniques are not widely used, due to the problems in maintaining feedback and relatively low spatial resolution. To prevent the "jump to contact", magnetically controlled feedback has been developed for accurate registration of reversible force-distance curves near the inflection point in the energy potential [97]. Another development uses a tunnelling current feedback to control a smallest possible tip-sample separation and thus achieve the true atomic resolution [98]. An electric field has been also employed for imaging liquid films and droplets [99]. Recently, a so-called phase-lock technique was proposed for stabilising the cantilever resonance when approaching the surface [100]. All these recent advances give a certain hope that non-contact SFM techniques will find a wider range of applications in the near future.

2.2.1.3
Intermittent Contact

In this mode, the cantilever oscillates near its free resonant frequency ω_0 with an amplitude which is sufficiently high to overcome adhesion forces [101]. The tip rapidly moves in and out of the sample surface so that it stays in contact only for a short fraction of the oscillation period. Depending on the cantilever type, the frequency typically varies from 50 to 500 kHz, and the amplitude changes up to 100 nm.

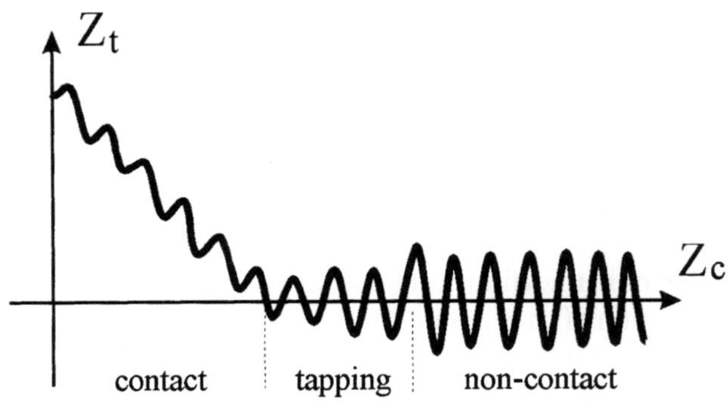

Fig. 10. Variation of the tip deflection Z_t as the oscillating cantilever approaches the sample surface. Z_c is the distance between the cantilever base and the surface

Figure 10 outlines the variation of the tip deflection as the oscillating cantilever is brought towards the surface. Depending on the tip-surface separation, three regimes can be distinguished: (i) non-contact, (ii) intermittent contact (IC), or tapping, and (iii) contact. At large distances, the cantilever oscillates with a constant amplitude which is determined by viscous damping of the cantilever body in air. When approaching the surface, the oscillations are amplified by the positive force gradient as it reduces the effective spring constant of the cantilever. Eventually, the oscillations become unstable and the tip jumps into contact. Still the energy of the cantilever is high enough to overcome adhesion forces and to be detached from the surface. The amplitude and the contact duration increase with decreasing the distance between the surface and the cantilever base. At some point, the amplitude drops rapidly and the tip become trapped by the sample. Beyond this point, the tip stays permanently in contact with the surface.

In order to understand the origin of the amplitude and phase images in taping mode SFM, it is instructive to examine the cantilever response as a function of the distance and frequency at a fixed lateral position. Figure 11 depicts variation of the amplitude, A_0, and the phase shift, φ, as a function of time (Fig. 11a), the oscillation frequency ω (Fig. 11b) and the cantilever-base displacement Z_c (Fig. 11c–f).

The diagrams in Fig. 11b can be obtained by the so-called "frequency-sweep" method, where the lateral position and the distance Z_c are fixed, while the frequency is varied around ω_0. The Z_c value corresponds to a given set-point ratio of the amplitude in contact to the free amplitude, $r_{sp}=A_{sp}/A_f$. Depending on the tip-sample interaction, both the amplitude and the phase curve shifts in a certain direction. When the overall force is repulsive, the resonance frequency moves to higher values and results in a positive phase shift $\Delta\varphi=90°-\varphi>0$, where the phase shift of 90 ° corresponds to the free cantilever oscillations at $k_s=0$ in Eq. 12. When the force is attractive the resonance frequency decreases compared to the free cantilever and $\Delta\varphi$ becomes negative. The situation in Fig. 11b corre-

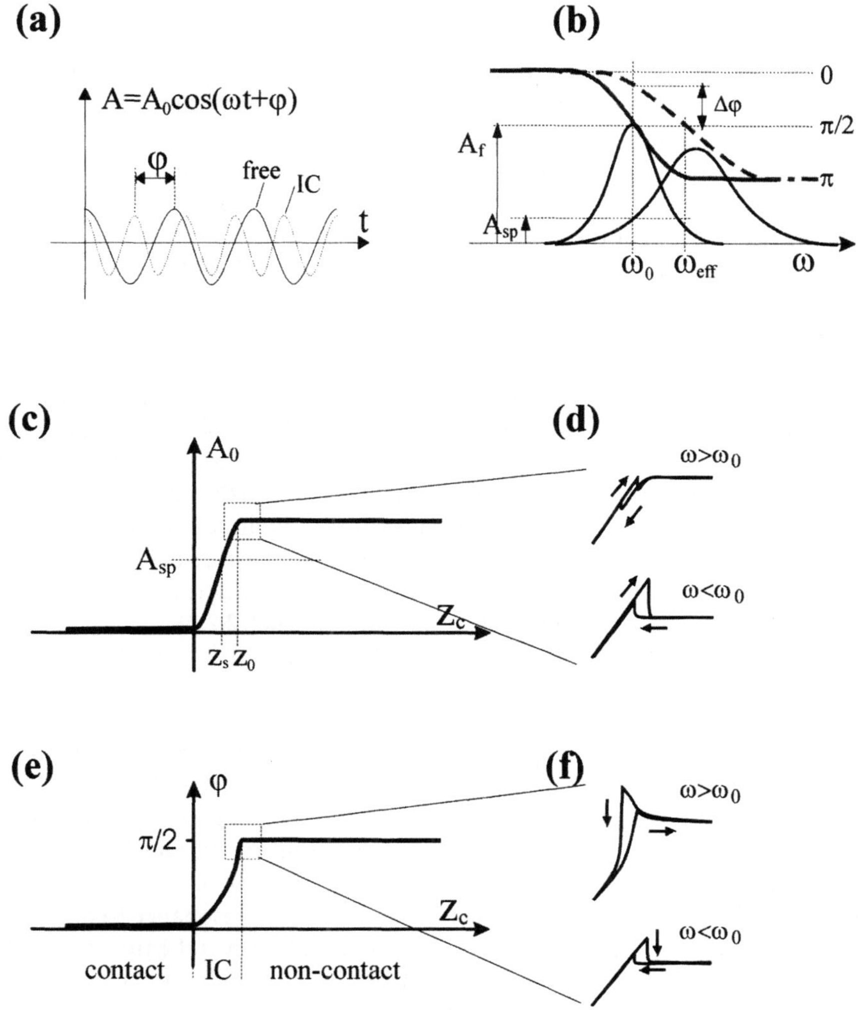

Fig. 11. a Tip oscillations at $\omega=\omega_0$ for the free cantilever and during intermittent contact (IC) with the sample, **b** Variation of the amplitude and the phase versus frequency obtained by sweeping the frequency around $\omega=\omega_0$; Variation of the **c** amplitude and **e** phase versus tip-sample separation as the tip goes from non-contact to the contact range via the IC regime. The curves on the right side of **c** and **e** display amplitude-distance **d** and phase-distance **f** variation recorded while approaching and retracting the cantilever near the onset of the IC repulsive range (squares in (**c**) and (**e**)). The curves were recorded above and below the resonance frequency ω_0 of the free cantilever. Reproduced after [116]

sponds to the repulsive interaction. The amplitude peak retains its symmetrical shape and becomes broader. Note that the amplitude peak can occur at frequencies which are higher than the resonance frequency of the damped cantilever. Such changes are typically observed for soft polymers [102]. The broadening indicates the decrease of the quality factor, caused by the energy dissipation due to the tip-sample interaction, whereas the frequency deviation is consistent with the sample oscillation [103]. For hard samples, such as glassy polymers, one can observe truncation of the amplitude peak and hysteresis of the amplitude and phase curves which are characteristic for a strong repulsion [104–106]. Under certain conditions, the system can also reveal bistability, where two stable states with different amplitudes occur at the same frequency [107]. This can result in flipping or inversion of the phase contrast.

The diagrams in Fig. 11c-f can be measured by the "force probe" method, when the amplitude and phase are measured as the tip approaches and retracts the surface vertically. In the non-contact range, both the amplitude and the phase retain their constant values (Fig. 11c,e). When the tip enters the intermittent contact range ($Z<Z_0$), both the amplitude and the phase reduces almost linearly on approaching the surface. The deviation of the amplitude signal from a certain set-point value A_s is used by a feedback loop to maintain the separation Z_c between the tip and sample constant, and hereby visualise the surface structure. When the surface composition is uniform, the amplitude variation is mainly caused by the surface topography. However, if the surface is heterogeneous, the variation in the amplitude can be affected by local differences in viscoelasticity [108–110] and adhesion [111] of the sample (Sect. 2.2.2).

In the transition region between the attractive and intermittent contact (Fig. 11d,f), the cantilever response depends on the oscillation frequency, ω, with respect to the resonance frequency of the free cantilever, ω_0 [112, 113]. At distances greater that Z_0, the positive force gradient leads to a reduction of the resonance frequency. When $\omega \geq \omega_0$, the lower frequency shift results in a decrease of the amplitude (Fig. 11d). In this distance/frequency range, the microscope can be operated in a purely attractive field preventing the contact. When the tip reaches the repulsive part of the interaction potential, the negative force gradient increases the effective spring constant and shifts the resonance frequency to higher values. The resulting increase in the amplitude moves the tip towards the repulsive field which causes a further increase in the amplitude. This auto-amplification process makes the dynamic system unstable until it reaches the IC regime at $Z<Z_0$. A different behaviour is observed when scanning below the resonance frequency ($\omega<\omega_0$). In the attractive force region, the amplitude increases until it becomes equal to the mean distance between the tip and the surface. At this moment, the tip snap into contact and enter the IC range. In contrast to the $\omega \geq \omega_0$ regime, it is hardly possible to realise a stable non-contact operation as the amplitude is amplified on approaching the surface. Compared to the amplitude, the phase (Fig. 11f) is more sensitive to the sign of the forces and to the tip-surface distance. In the transition range, inversion of the phase contrast might occur upon variation of the free amplitude or the set-point ratio [109–112].

For many reasons, the tapping mode is one of the most versatile developments in scanning force microscopy [114]. It was originally introduced for stable imaging at ambient conditions as the probe penetrates the contamination layer rapidly and intermittently. Besides that, there are three other key advantages of the tapping mode over conventional contact SFM. Forces exerted on the sample can be minimised due to the enhanced sensitivity towards small variations in amplitude at the resonant conditions. Thus, the tapping mode becomes applicable for imaging soft and fragile samples (Sects. 3 and 4). Secondly, there are no lateral forces that can tear the sample or damage the tip because the tip only contacts the surface briefly during each oscillation. This property is of primary importance for imaging of objects that do not adhere strongly to the substrates. Finally, the cantilever oscillates at high frequencies from 50 to 500 kHz, where many viscoelastic materials, such as polymers, respond elastically.

2.2.2 *Scanning Force Microscopes for Studying Material Properties*

A scanning force microscope probes a number of material properties such as elastic modulus, viscosity, friction coefficient as well as the work of adhesion and the surface energy by sensing different forces exerted on a sharp tip in the near field of the surface. The local mechanical properties are of great importance for designing microscopic devices consisting of moving parts that experience intermittent deformations ranging from 10–100 nm and local forces as low as nanoNewtons [115]. Also development of new functional materials and nanocomposites requires information about interfacial interactions and mechanical properties on the nanoscale [20, 23]. In conventional SFM experiments, these properties contribute equally to the net force (Sect. 2.1). In order to analyse each contribution separately, the experiment should be specially designed where one defines the mechanical field, the frequency range, the tip geometry, and the cantilever spring-constant.

2.2.2.1 Quasi-Static Acquisition of the Force-Distance Curves

At the beginning, the focus in using SFM for probing of micromechanical properties has been placed on elasticity to distinguish components with different stiffness. Somewhat latter, also viscoelastic and relaxation properties were investigated using modulation techniques. Figure 12 shows a typical forces-distance curve for a viscoelastic sample as obtained upon vertical displacement of the cantilever base towards and outwards the sample surface. A low frequency range around 1 Hz corresponds to quasi-static acquisition of the curve. The most strong changes in the cantilever response occur in the contact regime (B-C). Assuming that the spring constant, the position sensitive detector and the piezoelectric transducer are calibrated, one can measure the force F versus the sample deformation $\delta = Z_c - Z_t$. In the limit of small contact areas, the sample stiffness $k_s = F/\delta$ can be used as a measure of the Young's modulus E and the contact radius a related as $k_s \cong Ea$. However, one must take great care regarding the application of continuum theories for the contact mechanics in the limit of nanometer sized contacts and relaxation effects. The continuum approach was shown to provide a good basis for description of elastic contacts as low as 1–2 nm in radius [116].

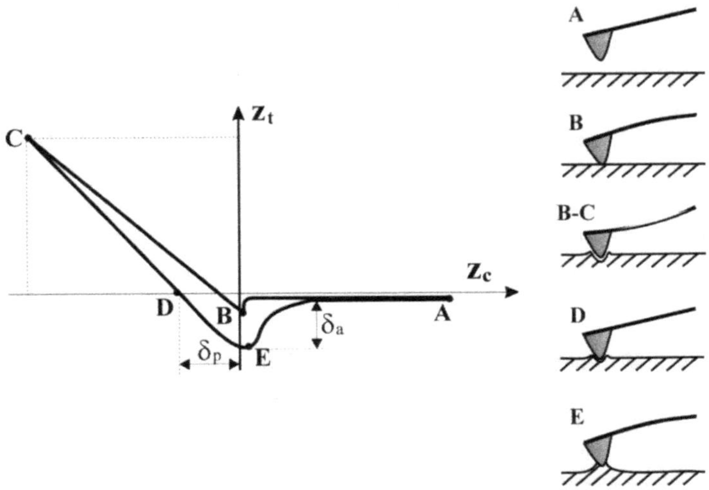

Fig. 12. A force-distance curve displays the vertical bending of the cantilever (Z_t) versus displacement of the cantilever base (Z_c) for a soft sample. The sample indentation $\delta=Z_c-Z_t$ as well as the plastic deformation δ_p at $F=kZ_t=0$ can be extracted from the curves recorded on approaching and retracting the cantilever

The unloading curve from point C to E gives the plastic deformation δ_p and the adhesion force $F_a=k_c\delta_a$ measured at a maximum downward deflection. The deflection after point D is indicative for the tip-sample adhesion, which can be caused by at least two effects: surface forces and stretching of a small portion of the sample adhering to the tip. In the latter case, the adhesion competes with cohesive forces. Thermodynamically, the stretching can be described by the balance between the work of adhesion and the cohesion energy, where the resulting force also depends on the stretching rate and the yield stress of the material [117]. Therefore, the "pull-off" force $F=k\delta_a$, which is often used to characterise the adhesive properties, might be of limited use for soft polymer samples.

Complementary to the normal force measurement, lateral forces can be recorded simultaneously to investigate friction properties of the materials. The point contact between the tip and sample is a suitable probe for studying the relation between friction and contact area as a function of the applied load and the surface composition [115]. The contact radius at a certain applied load can be found by measuring the contact stiffness, provided that the lateral force is smaller than the force of limiting friction [118, 119]. Depending on the surface deformation and material stiffness, an appropriate model has to be chosen for the calculation [120]. Tribological experiments are less reproducible on scanning of large areas because of the surface roughness, the chemical heterogeneity, and deformation of the surface asperities. To ensure the quantitative measurement, the LFM experiment has to be performed in ultra high vacuum using cantilever with a calibrated spring constant and a well characterised tip shape.

The quasi-static acquisition and interpretation of the force-distance curves is straightforward for elastic materials. The information is generally incomplete and less reproducible for polymers which demonstrate viscoelastic contact, plastic deformation, and ploughing type friction. Moreover, they exhibit a wide spectrum of relaxation times from 10^5 to 10^9 Hz [121].

2.2.2.2 Modulation Techniques

Modulation techniques offer the possibility to measure the elastic and retarded cantilever responses separately. In this case, either the sample or the tip is oscillated by an external means, such as piezoelectric transducer or magnetic field, in order to record the in-phase and out-off-phase amplitude signal. Figure 13a gives schematic representation of a modulation experiment where the amplitude, Z_t, and the phase difference, φ, are recorded while the z-position of the sample relatively to the cantilever base, Z_c, is modulated by a high-frequency transducer underneath the sample [122]. The relationship between the cantilever response and the material properties can be analysed in terms of the Voigt model consisting of two springs (k_s, k_c) and two dash pots (β_s, β_c) of the sample and the cantilever, respectively. For small excitation amplitudes, this model can be solved analytically assuming harmonic potentials, i.e. linear dependence of the interaction forces on the sample deformation $\delta = Z_c - Z_t$. In this case, the amplitude ratio and tangent of the phase difference between the sample displacement and the cantilever response are given by

$$\frac{Z_c}{Z_s} = \frac{k_s\sqrt{1+\left(2m\omega\beta_s / k_s\right)^2}}{\sqrt{(k_s + k_c - m\omega^2)^2 + 4m^2\omega^2\left(\beta_s + \beta_c\right)^2}} \tag{9}$$

$$\tan\varphi = \frac{2m\omega\left[\beta_s\left(m\omega^2 - k_c\right) + \beta_c k_s\right]}{k_s\left(k_s + k_c - m\omega^2\right) + \left(2m\omega\right)^2 \beta_s\left(\beta_s + \beta_c\right)} \tag{10}$$

Figure 13b depicts the frequency dependence of the cantilever amplitude for different k_s/k_c ratios. Depending on whether the tip oscillates below, near and above the contact resonance, the configuration of the experiment and the data interpretation can be different. Each of the modulation techniques has its advantages and limitations regarding derivation of the material properties from the experimental data.

The so-called Force Modulation Mode (FMM) was introduced in 1991 as a modification of contact-mode SFM [123] and has since found a wide range of applications in heterogeneous polymer systems [124–129]. The FMM imaging is performed by *z*-modulating the probe (or sample) at a low frequency of about 5 to 20 kHz and an amplitude greater than 10 nm. The frequency is sig-

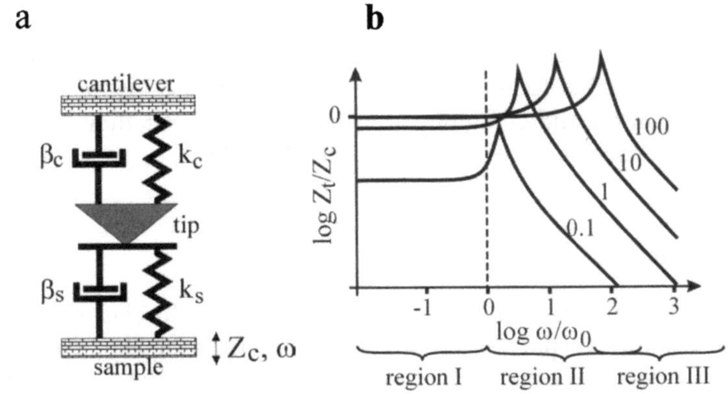

Fig. 13. **a** Rheological model for the cantilever response on applying a displacement modulation to a transducer underneath the sample; **b** the solution to the model gives the ratio of the amplitudes of the tip response Z_t to the sample excitation Z_c as a function of the $\log\omega/\omega_0$ for different ratios between the sample stiffness k_s and the spring constant of the cantilever k_c. Reproduced after [122]

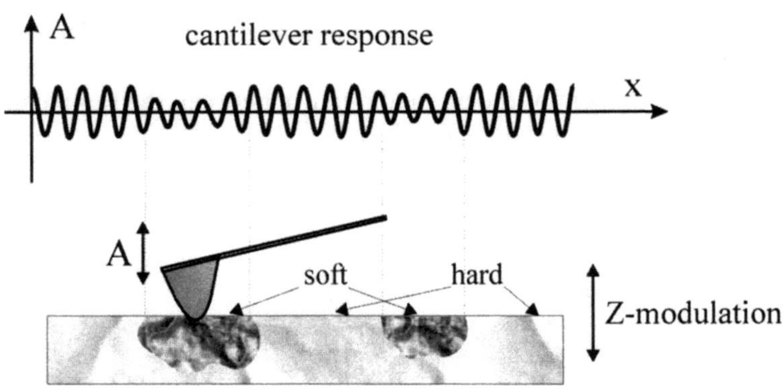

Fig. 14. Variation of the cantilever amplitude (*A*) recorded by force-modulation SFM where the z-position of the sample was modulated. The sample displays a heterogeneous surface consisting of soft and hard areas

nificantly faster than the scan rate, however well below the resonance frequency ω_0 of the cantilever (region I in Fig. 13b). To prevent tip-sample disconnection, the force on the sample is modulated in the repulsive range (Fig. 6). In this case, the tip does not vibrate, but rather indents the sample by imposing an oscillating force *F(z)* on the sample surface (Fig. 14). Under the same force, a stiff area will deform less than a soft area, and therefore leads to greater bending of the cantilever. Better control of the surface deformation can be achieved when the force is applied directly to the tip using e.g., a magnetic cantilever [130, 131].

Low-frequency acquisition of the curves corresponds to a non-inertial regime wherein the mass of the cantilever does not play any role and the system can be treated as two springs in series. The in-phase and out-of-phase mechanical response of the cantilever in FMM-SFM was interpreted in terms of stiffness and damping properties of the sample, respectively [125, 126]. This interpretation works rather good for compliant materials, but can be problematic for stiff samples. Assuming low damping, the cantilever response (Eqs. 9 and 10) below the resonance frequency ω_0 for the case of is given by

$$Z_t/Z_c \to k_s/(k_s+k_c)$$

$$\tan\varphi \to 0$$

For stiff samples ($k_c << k_s$), the amplitude ratio is $Z_t/Z_c \approx 1$ and the deflection of the cantilever almost does not depend on the stiffness (Fig. 13b). Under this condition, FMM-SFM can hardly distinguish surface areas differing in stiffness. To overcome this limitation and obtain measurable changes in the deflection, one has to use either stiff cantilevers or large amplitudes. These may cause lateral sliding of the tip, and thereby diminish resolution and incorporate frictional effects into the image [131, 132].

Several techniques were developed to study local viscoelastic properties at the contact area without scanning. In this case, one loses the lateral mapping but gains the possibility to perform force modulation spectroscopy at different frequencies ranging from 50 Hz to 10 kHz. The measurements can be done under normal [133–135] as well as shear [136] loading using a standard scanning force microscope. Irrespective of the loading conditions, the sample is oscillated at a very small amplitude, typically of 1–2 Å, to reduce microslip. The normal force modulation data are more complicated to interpret compared to the shear modulation spectroscopy, as the contact area is not constant and the cantilever response contains always a lateral movement due to the cantilever bending [132, 137]. These techniques have been used for investigation of the surface viscoelasticity of polystyrene films as a function of molecular weight [133], and to measure the contact area creep and relaxation times of poly(vinylethylene) [136].

To minimise the friction effect, it has been proposed to use smaller amplitudes and higher frequencies [122, 137]. The so-called scanning local-acceleration microscopy (SLAM) is another modification of contact-mode SFM which was implemented by vibrating the sample at a frequency above the highest tip-sample resonance (region III in Fig. 13b). In this frequency range (around 1 MHz), the cantilever response to the sample excitations becomes independent of the cantilever stiffness and depends linearly on the "contact stiffness" and reciprocally on the cantilever mass m (Fig. 13b):

$$Z_t/Z_c \to k_s/m\omega^2$$

$$\tan\delta = {}^{2m\omega\beta_s}\!/_{k_s}$$

The linear dependence on k_s is a strong advantage compared to the force-modulation SFM. Due to the inertia of the cantilever mass at high frequencies, the tip cannot follow completely the displacement of the sample. This results in smaller elastic deformations and low forces in the μN range, which are measured by *accelerated* mass md^2z/dt^2. Recently, the SLAM technique has been advanced toward variable temperature experiments [138]. Before this development, temperature ramps have been used to perform local calorimetry [139].

Information about the local mechanical properties can be also obtained by choosing the frequency close to the resonance of the tip-sample assembly (region II in Fig. 13b). Under these conditions, both the amplitude and the phase respond strongly to small variations in the stiffness and damping properties of the sample. This approach was realised in intermittent contact SFM which measures the variation of the amplitude, $\Delta A = A_f - A_{IC}$, and the phase shift, $\Delta\varphi = \varphi_f - \varphi_{IC}$, between the free and intermittently interacting cantilever (Fig. 11). Experimentally, the amplitude is maintained constant at a certain set-point value A_s by a feedback circuit in order to generate topographic image, while the phase shift is used to map the physical properties of the surface.

In spite of the apparent sensitivity to the material properties, the direct assignment of the phase contrast to variation in the chemical composition or a specific property of the surface is hardly possible. Considerable difficulties for theoretical examination of the tapping mode result from several factors: (i) the abrupt transition from an attractive force regime to strong repulsion which acts for a short moment of the oscillation period, (ii) localisation of the tip-sample interaction in a nanoscopic contact area, (iii) the non-linear variation of both attractive forces and mechanical compliance in the repulsive regime, and (iv) the interdependence of the material properties (viscoelasticity, adhesion, friction) and scanning parameters (amplitude, frequency, cantilever position). The interpretation of the phase and amplitude images becomes especially intricate for viscoelastic polymers.

An approximate solution of the equation of motion can be found assuming harmonic oscillations even though an intermittent and non-linear force dependence is introduced [102, 140–143]. For high oscillation frequencies in a MHz range, where most of the polymer samples behave elastically (low damping, high stiffness), the relation between the cantilever response and the material properties is given by

$$\frac{A_{IC}}{A_f} = \frac{\sqrt{(\omega_0^2 - \omega^2)^2 + 4\beta_c^2\omega^2}}{\sqrt{(\omega_e^2 - \omega^2)^2 + 4\beta_e^2\omega^2}} \to \frac{1}{Q_c}\frac{k_c}{k_s} \tag{11}$$

$$\tan\varphi = \frac{2\beta_e\omega}{\omega_e^2 - \omega^2} \to \frac{2\beta_e m\omega}{k_s} \tag{12}$$

where $Q_c = \omega_0/2\beta_c$ – quality factor of the free cantilever, $\omega_e = \omega_0 + \Delta\omega$ and β_e – effective resonance frequency and damping coefficient of the tapping cantilever, respectively.

At high set-point amplitude ratios $r_{sp}=A_s/A_f$, when the sample deformation and hence k_s are relatively low (light tapping), the phase shift at $\omega=\omega_0$ can be written as

$$\Delta\varphi = \frac{\pi}{2} - \tan^{-1}\frac{2\beta_e m\omega_0}{k_s} \approx \frac{k_s}{k_c}Q_e \tag{13}$$

where $\varphi_f=\pi/2$ is a maximum shift due to the viscous damping in air, $Q_e=\omega_e/2\beta_e$ – effective quality factor of the tapping cantilever assuming that $\omega_0>>\omega_e-\omega_0$. Therefore, at light tapping of an elastic sample $Q_e\approx Q_c$.

These equations lead to a few important consequences. (i) Since both r_{sp} and $\Delta\varphi$ depends linearly on k_s, the phase signal will not distinguish between two elastic areas differing in stiffness under the condition of the constant set-point ratio ($A_{IC}=A_s$). (ii) The phase contrast may arise from differences in elastic properties once the sample provides a channel for energy dissipation ($Q_e<Q_c$). In this case, the sample stiffness affects the phase signal indirectly as it determines the contact areas and the duration of the contact. (iii) The formula for $\Delta\varphi$ predicts a negative shift, i.e. a phase lag greater than 90°, when the overall force acting on the tip is attractive (k_s, $\omega_e-\omega_0<0$). In contrast, the phase shift is positive when the overall force is repulsive (k_s, $\omega_e-\omega_0>0$). To summarise, stiffer and less viscous areas (large k_s and Q_e in Eq. 13) will be imaged with a greater $\Delta\varphi$ corresponding to a lower level of energy dissipation.

Amongst different channels of the energy dissipation, the most relevant are the sample viscosity, the contact friction, and the adhesion hysteresis. When the drive frequency is chosen to be ω_0, the power that is dissipated when the tip periodically interacts with the sample can be written as

$$P = \frac{1}{2}\frac{kA^2\omega_0}{Q_c}\left[\frac{A_f}{A_s}\sin\varphi - 1\right]$$

Complementary to model calculations of the non-linear equation of motion [112, 113, 140–143], this approach is based on analysis of the energy flow in and out of the dynamic system [144]. The equation does not specify the source of damping and simply shows how to interpret phase images taken on resonance at constant amplitude. As long the microscope operates either in attractive or repulsive regime, then any changes in the phase signal are caused by the energy being dissipated in the tip-sample junction, and $\sin\varphi$ is proportional to the dissipation power *P*. In non-contact microscopies, it can be convenient to feedback on the frequency to maintain the lever on resonance. In that case, the phase is held constant and changes in *P* appear in the cantilever amplitude [145].

The experiment shows that under the typical tapping conditions, the phase signal is mainly caused by the repulsive forces due to the sample deformation [102]. For example, for a polymer blend consisting of polypropylene (PP) phase embedded in a polyurethane (PU) matrix, the stiffer PP demonstrated a maximum power dissipation of about 0.8 pW, which corresponds to an energy dissipation of 66 eV per tap [146]. The softer PU exhibited an energy dissipation which was about three times higher, i.e. 202 eV per tap. The latter is equivalent to 8×10^3 k_BT (at 293 K), which roughly corresponds to a few thousands of atoms involved into the tip-sample interaction. For the interaction volume V=100 nm^3 and the tip radius R=10 nm, one can estimate a contact radius to be about 5 nm assuming the elastic contact with $V\approx a^4/R$.

The one exception in which phase contrast is not due to the dissipation arises when the tip jumps between attraction phases (>90°) and repulsion phases (<90°). Since sine is a symmetric function about 90°, the phase changes symmetric even if there are no losses in the tip-sample interaction. The relative contribution of the repulsive and attractive forces can be estimated experimentally from the frequency-sweep curves in Fig. 11b by measuring the effective quality factor as $Q_e=\omega_0/\Delta\omega_{1/2}$, where $\Delta\omega_{1/2}$ is the half-width of the amplitude curve. The relative contribution of the attractive forces was shown to increase with increasing the set-point ratio $r_{sp}=A_s/A_f$. Eventually, this may lead to the inversion of the phase contrast when the overall force becomes attractive [110, 112]. The effect of the attractive forces becomes especially prominent for dull tips due to the larger contact area [147].

In conclusion, tapping mode SFM allows rather gentle monitoring of the surface topography and, at the same time, discriminates structural details differing in their viscoelastic properties and adhesion. It must however be emphasised that the influence of the sample material properties on the cantilever response are intricate. Neither are topographic tapping images independent of sample properties, nor is it possible to extract material properties readily from the tapping data. Moreover, interpretation of the images requires the contact stiffness, which is connected to the elastic modulus through the contact area. Since the contact radius is usually not known, one should be careful in the interpretation of the stiffness and phase images. At least, accurate evaluation of the tip geometry is necessary to obtain a reliable value of the elastic modulus (see Sect. 2.3.1).

2.2.2.3
Chemical Force Microscopy

Lately one has been able to encounter experimental studies more frequently denoted "Chemical Force Microscopy", CMF. This includes various attempts to observe tip-surface interactions which are specific to the chemical constitution of the surface. Mostly, CFM involves modification of the tip by a surface layer with molecules which contain particular functional groups, i.e. hydrophilic or hydrophobic moieties, hydrogen bonding groups, ionic substituents and molecular units which can undergo electron-donor-acceptor interactions. However, sometimes the term "Chemical Force Microscopy" is just used for any method which can provide a material specific contrast. Depending on the specificity, CFM provides valuable information on the nanoscale composition complementary to other surface characterisation methods which are sensitive to the chemical constitution, i.e. vibration spectroscopy, SIMS, XPS, energy dispersion X-ray analysis and many others [148]. The intriguing aspect is that investigations can be performed under ambient conditions and, most important, that the relevant interactions are probed locally on the scale of a few nanometers.

For example, chemical contrast images were obtained by lateral force microscopy (LFM) from a topologically flat surface of a self assembled monolayer consisting of chemically different domains. In order to make the "chemical" adhesion the dominant contribution to the friction signal, the tip was modified by a monolayer with appropriate terminal groups [149–155]. However, since LFM operates in contact mode, the surface deformation is inevitable.

Non-contact and intermittent contact methods were introduced to overcome the effect of sample deformation in contact mode imaging. Depending on the tip-surface separation, the phase shift between free oscillation in air and oscillation in the near field of the surface provides information either about the attractive force gradient [112] or the contact adhesion. It has been demonstrated that the phase contrast in tapping mode imaging of chemically distinct monolayer regions and the adhesion forces are directly correlated, i.e. increasing adhesion forces lead to an increase in the phase lag [111]. Under reasonable experimental conditions, a change of 5 nN in adhesion force can produce a phase shift of about 2° [142]. However, as discussed above, the net phase signal can be further affected by the capillary condensation in the contact area and viscoelastic properties of the sample.

Chemical interactions between the tip and the surface can be determined also from the force-distance curves. The so-called "pull-off force", which is associated to adhesion of the tip towards the sample, can be scanned directly over the X-Y plane [156]. In the case of conventional data sampling, image resolution is limited because of the large amount of data which have to be collected for each image point [129]. High sampling rates can be achieved by means of the so-called Pulse Force Mode (Fig. 15), which is based on analogous processing of the force-distance curves [157]. PFM-SFM is able to measure simultaneously the local stiffness and adhesion at the surface. A sinusoidal modulation is applied to the z-piezoelectric scanner and only a few characteristic points on the force-distance curve are recorded as the sample surface and the probing tip are brought into cyclic contact. For example, the inflection point with a maximum dF/dz slope can be taken for the sample elasticity and the "pull-off force" is characteristic for adhesion.

The modulation frequency is typically in the range from 100 Hz to 3 kHz, and thus much lower than the resonance frequencies of the cantilever and the scanner. This enables better control of the forces exerted on the sample. The z-modulation amplitude can be varied between 10 nm and 1 μm to ensure that that the tip is retracted from the surface. Shear forces are reduced permitting investigation of soft samples because of the small duration of the tip-surface contact, between 10^{-3} and 10^{-4} s. Pulse force mode SFM has been used to map adhesion of heterogeneous polymers in dependence of temperature and molecular weight as well as map electrostatic double-layer interactions [158–160].

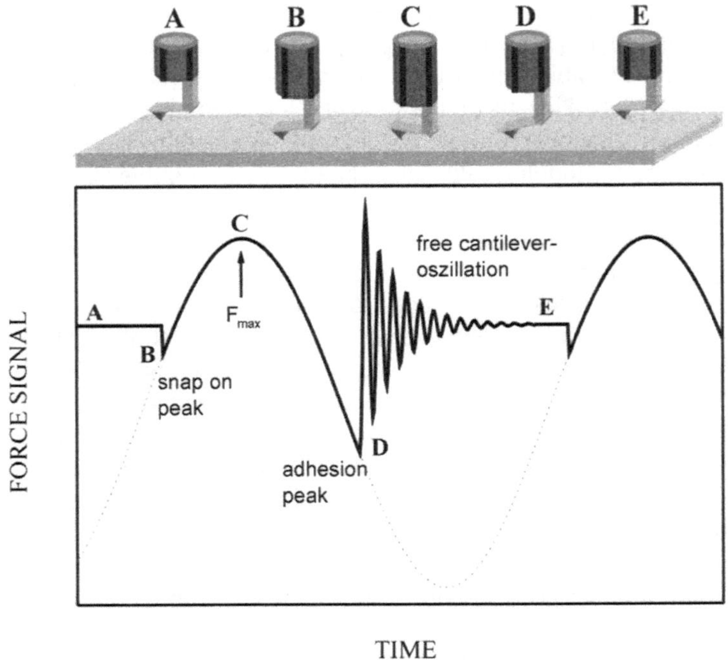

Fig. 15. Schematic illustration of the principle of pulse mode SFM [157]. The schematic illustrations of the tip-cantilever assemblies above the force diagram correspond to different positions of the cantilever base relatively to the sample surface

Another approach to extract the adhesion force has been demonstrated recently by Burnham et al. using scanning local acceleration microscopy (SLAM) [122]. If a high frequency oscillation with a modulated amplitude is applied to the sample, the cantilever can respond nonlinearly. Similar to the non-linear behaviour of an electrical diode, the cantilever signal exhibits changes above a certain threshold amplitude. Both the signal height and the threshold amplitude depend on the elastic modulus and the surface energy of the sample. Recently, a principally new approach to sensing of the chemical composition has been demonstrated by using SFM tip as a scattering centre for infrared radiation in the near field of the sample [160]. This aperture-less technique enabled measurement of IR-adsorption spectra with resolution below $\lambda/100$, i.e. on a scale of about 100 nm.

2.2.2.4
Other Modes and Scanning Probe Techniques Useful for Polymers

Different SPM systems were developed to study the thermal properties. Thus a tiny thermocouple can be used to measure the heat flow from the surface and to test the local thermoconductivity of polymer surfaces [161]. Recently, a bimetallic cantilever has been used as temperature sensor to investigate phase transitions of n-alkanes with a heat sensitivity of 500 pJ for a sample mass as low as to

7 pg [162]. Also micromechanical thermogravimetry could be performed using a piezoresistive cantilever enabling simultaneous control of the temperature and nanogram quantities of material [163]. Based on a conventional STM, a scanning thermopower microscope (STPM) was developed to measure simultaneously the topography and the heat flow of an organic monolayer. The thermopower signal is not sensitive to the surface roughness, but it permits imaging of single molecules and provides information on the chemical potential of the adsorbate [164].

An interesting approach to the investigation of cell membranes was introduced by the so-called patch clamp technique (PCT) based on SFM [165]. A patch of the membrane of a biological cell was fixed at the end of a pipette and studied by SFM. The cell can be kept alive for several days, while one can investigate both morphology and micromechanical properties of the cell membranes on the nanometer range. Structures as small as 10–20 nm and forces as low as 0.1 nN could be resolved which enabled investigation of the binding of labelled antibodies, pore formation, and dynamics of the membrane structure.

Special experiments were designed to measure long range electrostatic and magnetic forces. The so-called "lift mode" uses a combination of contact and non-contact imaging in order to determinate variations in the surface topography and electromagnetic surface effects [166]. The same surface area is scanned two times. First, the tip scans in a close proximity to the surface to record the topography. Afterwards, the tip is lifted to a larger separation and scan the surface once again along the topographic profile recorded previously in order to obtain only the force micrograph without variations caused by the surface roughness.

Recently, pneumatic SFM has been developed for investigation of porous materials regarding pore size, distribution of the pores, and the gas permeability [167]. Figure 16a shows a schematic diagram of the pressure chamber which is sealed hermetically with the membrane to be investigated. The pore structure was identified as the gas (N_2) started to stream through the pores resulting in the upward deflection of the cantilever away from the membrane surface. Figure 16b shows a pneumatic SFM image of a nuclear pore membrane of polycarbonate (pore diameter=400 nm) at a pressure difference of ca. 0.4 bar. The observed elevations corresponds to the gas cones from the tiny nitrogen streams through the pores.

2.2.3
From Ambient Conditions to Measurement in Liquid and Vacuum

2.2.3.1
Ambient Conditions

At ambient conditions the sample surface is always covered by a thin contamination layer. This layer is mainly composed of water, however, also other compounds can adsorb to the sample surface and the surface of the tip. Depending on the vapour pressure of the contaminant, the layer thickness can range from 2

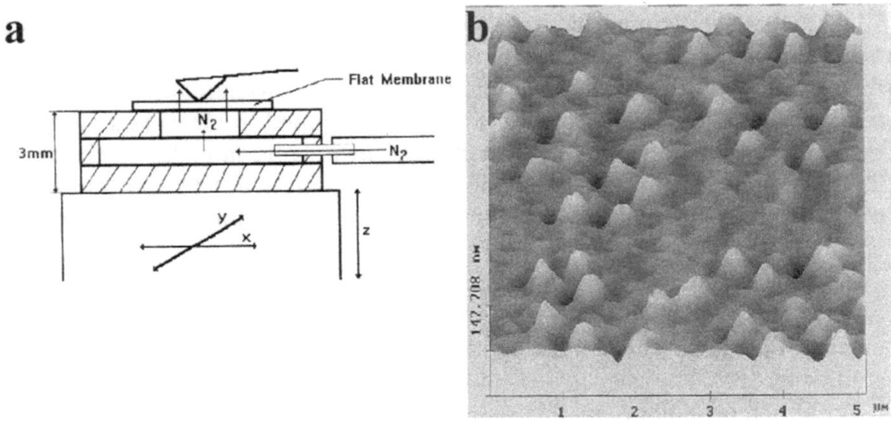

Fig. 16. a Schematic diagram of the pressure chamber for characterisation of flat membranes by means of a pneumatic SFM. **b** Height image of a nuclear pore membrane of polycarbonate recorded by the pneumatic SFM at a pressure difference of 0.4 bar. The elevations are resulted from the nitrogen streams through the pores. Reproduced from [167]

to 50 Å. For example, at 40–50% relative humidity, mica is covered by a water layer with an average thickness of 2 Å [168, 169]. At higher humidity, even multilayers are formed.

As the tip of the probe comes into contact with the contamination layer, capillary action causes formation of a meniscus around the tip which gives an additional contribution to the interaction between the probe and sample (Fig. 3b). Depending on the wetting conditions, the probe can be either pulled to the surface or repelled. Usually the capillary forces are attractive as the regular Si and Si_3N_4 probes are wettable by water and organic molecules. The magnitude of the capillary force depends on the radius of curvature and the opening angle of the tip [170]. Formation of a liquid neck as well as the force exerted on the tip can be detected by SFM operating near its resonance frequency [171].

2.2.3.2 Ultrahigh Vacuum SFM

Contact SFM can be performed at much lower forces in vacuum than in air. Due to elimination of the thin water layer, also operation in non-contact mode became much more stable at distances near to the surface. UHV-SFM in contact mode allows easily nanometer resolution and even atomic-scale features could be resolved by non-contact SFM using a special feedback scheme [98, 172]. Thus, UHV conditions strongly advance non-destructive imaging of soft surfaces and make the measurements more reproducible and quantitative.

The severe limitations are high costs of the UHV set up and special requirements to samples that should sustain the UHV conditions. The first UHV-AFM with atomic resolution at 5 K was demonstrated in 1992 using a tunnelling sensor to measure the cantilever deflection [173]. Other groups developed alternative designs based on sensing capacitance, interferometric sensors and beam deflection [174–176]. Further improvements in this field are connected with microfabrication of piezoresistive cantilevers [172].

2.2.3.3 Measurement in Liquid

One of the main advantages of scanning force microscopy is based on its ability to perform in situ imaging in liquids, something not possible with electron microscopy. Interfacial forces and morphology variation at the liquid/solid interface can be investigated as a function of the solvent quality, the ionic strength, and pH [177–181]. Even structural forces indicating layering of the solvent between the tip and sample could be identified [182, 183]. In addition, the immersion of the tip into the liquid prevents capillary condensation (Eq. 4) and may cause disjoining pressure (negative A in Eq. 5). Consequently, the repulsive force exerted on the sample can be set lower than in air, typically about 1 nN, so that soft polymers and biological samples can be imaged without distortion of their native structure. By imaging a low adhesion surface under water, Binnig et al. were able to bring an SFM tip into contact gently and achieve atomic resolution of the calcite surface [184].

The surface deformation could be reduced even further with intermittent contact SFM. Tapping mode imaging in liquids has been described by several groups [185–191]. The main focus has been put on biological systems such as DNA, cells, chromosomes and proteins. However, it turned out to be rather tricky to perform the measurements. The resonant frequencies were usually 2–5 times lower than in air and the resonant peak gets strongly dampened and broadened [192, 193]. Because of acoustic excitation of the cantilever holder and the body of fluid, the spectrum can be superimposed by other resonance's which not sensitive to the surface approach and cannot be used for the feedback control [185].

Better control of the cantilever oscillation in liquid environment can be achieved when the cantilever is oscillated directly by an external force. This idea was implemented by the so-called Magnetic-Alternative-Current Mode (MAC Mode) [194]. A magnetic cantilever is driven by an external magnetic field which is generated by a solenoid placed beneath the sample. The direct excitation of the cantilever avoids unwanted resonance's from the cantilever holder, the fluid body, and the sample itself. Furthermore, the improved signal-to-noise ratio allows smaller oscillation amplitudes and set point ratios A_{sp}/A_f closer to 1. Both factors result in a significant reduction in the energy deposited into the sample, which can be estimated as $E \approx \frac{1}{2}k\left(A_f^2 - A_{sp}^2\right)$. Magnetic probes were used for the determination of the sample stiffness [187, 195] as well as the local compliance of ordered liquid layers at solid-liquid interfaces [183, 196].

It must be noted that also the design and operation of a fluid cell is of great importance. The cell should be inert against the liquid and sealed in order to avoid leakage or evaporation of the liquid. Any variation of the total mass changes the resonant properties of the cantilever assembly. It can disturb the measurements or even result in unstable imaging conditions. Both the liquid and the sample surface should be of maximum purity. Adsorption of some contaminant on the surface of the cantilever or the probe tip has to be taken into account as it can result in imaging artefacts and even destruction of the force sensor. The description of different designs of fluid cells can be found elsewhere [57].

To summarise this section, today's scanning force microscopy involves a large variety of methods differing in the probed force field (contact, non-contact, intermittent contact), operating conditions (from vacuum to condensed liquid) and registration principles (static, dynamic). Advanced scanning force microscopes often combine the different methods. The most intriguing advantages of SFM over conventional microscopy techniques, apart from the superior resolution and the operation under various environmental conditions, result from the ability to measure micromechanical properties, interfacial forces and to manipulate of the surface structures on the nanometer scale (Sect. 3).

2.2.4
Improved Force Probes

The so-called optical lever [86, 87] is widely used to sense the forces between the tip and sample down to 1 pN (Fig. 5). In this detection system, vertical displacements ΔZ_t of the tip are converted to angular deflections of a laser beam reflected off the back side of the cantilever into a position sensitive detector (PSD). The movement of the beam centre over the detector area is given by $\Delta X = g\Delta Z_t$. The optical lever acts as a displacement amplifier whose gain factor $g=d/l$ depends on the cantilever length l and the distance d between the cantilever end and the PSD centre. From the values $l=10^{-4}$ m and $d=10^{-2}$ m, for a typical microfabricated cantilever with an integrated tip we obtain the gain factor $g=100$, which can be even higher up to 1000 depending on the microscope construction. The angular sensitivity is limited by the laser power and noise in the system. Typically, tip displacements as low as 0.1 nm can be detected by the optical lever.

A suitable cantilever has to meet certain requirements regarding the spring constant and the resonance frequency [72]. In conventional contact-mode SFM, weak cantilevers with a low spring constant are recommended to decrease the sample deformation. On the other hand, thermal oscillations of the cantilever set a lower limit for the spring constant as $k_{min} \approx k_B T / z_t^2$. Imaging with a thermal noise below $z_t = 0.1$ Å, which is about 10% of the optical-lever sensitivity, would require a minimum spring constant of about 40 N/m. It is important to note that in contact-mode SFM the lower limit refers to the effective spring constant $k_{eff} = k_c + k_{int}$, which includes the cantilever spring constant k_c the interaction force gradient $k_{int} = dF/dz$ due to the sample deformation. Since both constants are interrelated via the contact area, these equations can be solved to determine the minimum value a_0 of the contact radius corresponding to k_{eff}=40 N/m. For small deformation, the interaction constant can be written as $k_{int} \approx E \times a_0$ and the spring constant can be determined from Eq. 6 as $k_c = K \cdot a_0^3 / \Delta \cdot R'$, where Δ is the sensitivity of the detection system and R is the tip radius. For Δ=1 Å and R=10 nm, a polymer sample with the elastic modulus K=1 GPa will undergo deformation with a contact radius a_0=4 nm. Therefore, the interaction constant of the soft sample will be only 4 N/m, which allows to reduce the spring constant of the cantilever to 36 N/m. Further reduction of the spring constant would require additional sample indentation to increase the contact area or measurement with a higher thermal noise.

Another requirement is that the resonance frequency of the cantilever has to be selected as high as possible to reduce the effect of acoustic and building vibrations. In order to sustain a high frequency, while reducing the spring constant, it is necessary to reduce the mass of the cantilever, i.e. the size and/or the material density should be chosen as small as possible.

Usually, microfabrication techniques are used to prepare cantilevers with integrated tips of various shapes, mass and spring constants [197, 198]. Depending on the cantilever geometry and material used to construct the cantilever [52], the frequency of commercial cantilevers typically varies from 15 kHz to more than 500 kHz, and the spring constants range from 0.01 to 100 N/m. Micromachining techniques can be used to prepare special probes such as meander-type cantilevers for bidirectional force microscopy [199].

A new type of cantilevers based on polymers has been microfabricated using multilayer thin film technology [200]. Novolak was used as a photoresist for microstructuring on Si or Si_3N_4 substrates with subsequent thermal curing. Tips were grown at the end of the cantilever using the electron beam deposition technique [201–204]. Due to the polymer structure, these cantilevers show very low force constants down to 0.003 N/m but still relatively high resonance frequencies up to 36 kHz.

Special cantilevers have been designed for in-vacuum as well as in-liquid operation. Piezoresistive cantilevers [205] and micromachined cantilevers with an integrated electronic sensor [206] make the deflection measurements more simple, however demonstrate lower force sensitivity.

2.2.4.1
Probe Geometry

Depending on the sample and the experiment to be done, tips of different sharpness, aspect ratio and chemical nature are required. Sharp tips are necessary to improve lateral resolution and reduce adhesion forces, whereas tips with a high aspect ratio, i.e. the ratio between the length and the diameter of the tip, are required for rough surfaces to record steep variations in height. Integrated tips of

Si_3N_4 or Si with a radius below 10 nm can be prepared using microfabrication techniques [197, 198]. The silicon nitride probe represents a pyramid with ca. 5 μm long sides of a square base and the opening angle of 70.5°. Etched silicon probes are conical in shape, with a base radius of 3–6 μm and a height of 10–20 μm. Compared to the pyramidal tips, Si probes have a higher aspect ratio and smaller opening angle.

Microfabricated SFM-probe tips are often blunt and have a wide opening angle. Electron-beam assisted chemical vapour deposition can be used to make tips with narrow opening angles and large aspect ratio [201–204]. The probes with a diamond-like composition are grown on top of standard pyramidal tips when the electron beam of a scanning electron microscope is focused onto the point of the pyramid provided that the vacuum chamber contains some organic contamination. By this deposition process one could make tips which are 1.5–2 μm long, with a radius of 10 nm and an aspect ratio >10. These tips have been successfully used for imaging DNA [207]. A combination of an atomic force microscope with a scanning electron microscope was proposed to shorten the time between growth and testing of the e-beam tips [208]. By changing the electron beam direction, it was also possible to create tips with hooked or ball-shaped ends which might be of interest for probing long range colloidal forces [178].

High aspect ratio SFM probes can be also made by a focused-ion-beam (FIB) technique [209–211]. Microtips were reproducibly grown up to 1.0 μm in length and 0.1 μm in diameter. A tip radius as low as 5 nm could be achieved and was found to degrade only slightly after extensive SFM imaging. Conventional tips can be modified by etching techniques so that a sharper probe apex is provided [212].

The mechanical deformation of the extra long tips limits the aspect ratio to a certain maximum. Also the probe apex can be damaged whenever the tip crashed into a hard surface and moves laterally. A more sophisticated but very promising approach to increase durability of the tips was introduced recently by Dai et al. [213]. Individual carbon nanotubes several nanometers in length and 5–20 nm in diameter were attached to silicon cantilevers. Due to the small diameter, the capillary forces could be reduced below 5 nN, permitting tapping-mode measurements with soft cantilevers (k=0.01 N/m) and a peak-to-peak amplitude of 10 nm. Because of their flexibility, the tips were resistant to damage from tip crashes, while their slenderness permitted imaging of sharp protrusions in surface topography. In addition, the nanotube tips can be used for STM imaging due to their electric conductivity.

Special probes were introduced to measure surface forces in colloidal systems as a function of the ionic strength and the concentration of surfactant molecules [214]. A so-called colloid probe can be prepared by gluing a silica sphere onto a conventional Si_3N_4 tip [179].

Over the last few years, chemical modification of the tips has attracted a lot of attention because of their application for chemical force microscopy. Different methods have been developed to functionalise a SFM-tip with either hydrophobic or hydrophilic molecules. Most of them are based on the monolayer self-assembly yielding a dense coating with a well defined surface structure [215, 216]. With modified tips one can easily identify patches differing in composition when studying a chemically heterogeneous surface [151, 152, 217–219]. This can be done either with lateral force microscopy or even by conventional SFM provided that the adhesion contribution to the net force is relatively large. Recently, a pure adhesive contrast has been achieved upon imaging a surface structured by the microcontact printing (μCP) technique [220] using a Pulse-Force-Mode [159] as well as tapping mode SFM's [111]. An important application of the modified tips is directed towards molecular recognition in biological systems [221, 222] and long range hydrophobic interactions [223].

2.3 Typical Problems Encountered when SFM Imaging

In this section, we will discuss a few selected problems which are often faced by the experimentalist studying polymer samples with SFM. To these problems we refer to the unknown shape of the probe apex (Sect. 2.3.1), deformation of the sample (Sect. 2.3.2), and limitations in lateral resolution (Sect. 2.3.3).

2.3.1 *Calibration of the Tip Shape*

The physical geometry of the tip is involved in the imaging process in many respects. In the contact regime, the tip radius determines the contact area which is directly involved in the adhesion and friction forces between the tip and sample. For easily damaged or weakly adherent samples, sharp tips are required to ensure scanning at lower forces with minimal sample damaging. In the non-contact regime, the finite size of the tip in combination with the relatively large gap increases the effective interaction area and therefore diminishes both lateral and vertical resolution of the images. Finally, the tip geometry is crucial for imaging of surface structures which are commensurate with the tip shape. Blunt tips, or tips with large opening angles, will cause steps to appear sloped, and corners or peaks to appear rounded (Fig. 7). Additionally, blunt tips are unable to reach down into narrow valleys to image the surface at the bottom of these valleys. These and other artefacts originating from convolution of the tip shape with the surface topography have been discussed elsewhere [224–226].

As was demonstrated theoretically, a single atom at the end of a rigid tip is not stable as well as tips with an infinite aspect ratio do not exist [227, 228]. Therefore, SFM tips are truncated and require a special procedure to calibrate their unknown geometry. Electron microscopy was widely employed for ex-situ examination of the tip shape [226, 229, 230]. This technique provides valuable information on the overall shape of the tip, however it is less suitable for characterisation of the nanometer sized probe apex.

Several authors proposed to use a specially designed substrate for in-situ calibration of the tip shape based on extraction of the tip geometry from the topographic profile of the substrate [231–236]. This procedure can be done before, during of or immediately after the experiment in order to monitor the degradation of the tip over extended use. Depending on the substrate structure, this approach can be used for microscopic characterisation of the overall shape of the tip as well as nanoscopic characterisation of the probe apex. In the former case, one can use various structures steeply protruding or sticking out of the surface. The high aspect ratio structures with sharp edges can be produced by different microfabrication techniques such as microlithography, ion-beam etching, or chemical vapour deposition [231–233]. An apparent disadvantage of such substrates is the intrinsic imperfection of the microfabricated structures. Actually, the substrates are prepared by the same techniques which are used for the tip fabrication, and therefore contain defects of the same size scale. Therefore, their application for testing of the outermost part of the tip is limited.

In order to measure the radius of the probe apex and identify atomic scale asperities on it, one has to use native structures such as colloidal particles and single crystals whose geometry is strictly defined. Figure 17 shows different substrates which can be used for calibration of probe apex. Spherical particles of gold with a diameter below 100 nm were used to calibrate the tip shape and evaluate the lateral dimensions of biomolecules [234]. For a parabolic tip in Fig. 17a the tip radius R is determined by the cut off height h_0 and the opening angle α as $R=h_0\sin\alpha/(1-\sin\alpha)$. When the particle radius r is larger than the tip curvature ($r>R$), the h_0 and α can be determine from the S-shape of the scan line:

$$\tan\alpha = dL/dZ$$

$$h_0=[L\cos\alpha - r(1+\sin\alpha)]/\sin\alpha,$$

where L is the lateral size of the particle deduced from the cross sectional profile [235].

When the particle is smaller than the tip curvature ($r<R$), the tip radius can be determined directly from the SFM profile as $R=L^2/4r$ (Fig. 17b). This calibra-

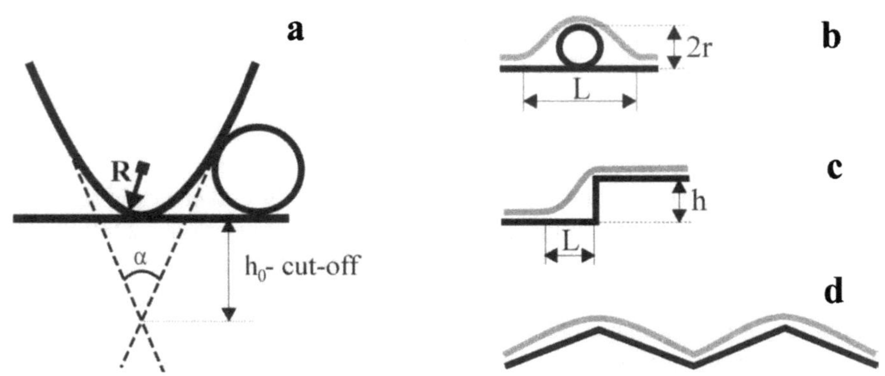

Fig. 17. Different methods for the shape calibration of the probe apex using **a** – a large sphere (r>R), **b** – a small sphere (r<R), **c** – a terrace of height h, **d** – a saw tooth surface of a single crystal formed of alternating crystalline planes

tion method works rather well if the particle has perfect spherical shape, is rigid, and do not move during the scanning process.

Crystalline steps enable very easy measurement the tip radius using the equation $R \approx L^2/2h$, where h is the step height (Fig. 17c). This calibration method requires a rigid substrate with many steps which are uniform in height in a range of 1–10 nm.

Single crystals with a stepped surface structure represent one of the most accurate calibration techniques (Fig. 17d). Such surfaces can be prepared by cutting a crystal at a certain angle relatively to the main crystallographic planes. Some of the cut planes are thermodynamically unfavourable resulting in the surface recrystallisation. For example, a saw tooth profile emerges on the surface of (305) cut crystal of $SrTiO_3$ [236]. The profile is formed of the alternating (101) and (103) planes inclined at 11 and 14 degrees with respect to the (305) surface of the wafer, respectively. In addition to the crystallographically defined surface structure, the wafer is rigid and does not change its structure during extended use at ambient conditions. This calibration method was used to quantify the contact area required for characterisation of micromechanical surface properties such as friction [237] and hardness [238].

2.3.2
Sample Deformation in Contact Mode SFM

For soft materials such as polymers and biomolecules, mechanical deformation of the sample caused by the tip is a point for serious concern. The consequences of the deformation are rather severe. First, it increases contact area and therefore reduces resolution. Second, the height profile of the surface can be underestimated. Third, the original structure of the sample can be destroyed irreversibly. Furthermore, the tip itself can be deformed when imaging hard samples [239].

There are two major sources of the deformation in contact-mode SFM: the elasticity of the cantilever and the adhesion between the tip and sample surface. For purely elastic deformation, a variety of models have been developed to calculate the contact area and sample indentation. The lower limit for the contact diameter and sample indentation can be determined based on the Hertz model without taking into account the surface interactions [79]. For two bodies, i.e. a spherical tip and an elastic half-space, pressed together by an external force F the contact radius a and the indentation depth δ are given by the following equations:

$$a=(3RF/4K)^{1/3}$$

$$\delta=a^2/R$$

where F is the applied force, and K is the elastic modulus of the material given by $K=2E/3(1-\nu^2)$ with E the Young's modulus and ν the Poisson's ratio. For a 10 nm tip, E=10 GPa and a typical SFM load of 1 nN, one can expect a contact radius and surface deformation larger than a=0.95 nm and δ=0.09 nm, respectively. From the contact area, the pressure can be estimated as $\sigma=F/\pi a^2=$ 335 MPa. This value is higher than compressive strength of polymers, which typically lies between 1 and 50 MPa, indicating that the microscope operates beyond the yielding point and may cause plastic deformation of the surface.

Also adhesion between the tip and sample can cause deformation of the sample. Several theories have been developed to include the effect of adhesive forces. In the JKR theory adhesion forces outside the contact area are neglected and elastic stresses at the contact line are infinite [80]. Even under zero load, the adhesion force results in a finite contact radius $a=(9\pi R^2\gamma/2\ E)^{1/3}$ as obtained from Eq. 7 for F=0. For example, for a tip radius R=10 nm, E=1GPa, typical surface energy for polymers γ=25 mN/m, and typical SFM load F=1 nN, the contact radius will be about a=9.5 nm and δ=9 nm, while under zero load the contact radius and the deformation become a=4.5 nm and δ=2 nm, respectively. The experiment shows that under zero load the contact radius for a 10 nm tungsten tip and an organic film in air is 2.4 nm [240]. The contact radius caused only by adhesion is almost five times larger than the Hertzian diameter calculated above. It means, that even at very small forces the surface deformation as well as the lateral resolution is determined by adhesion between the tip and sample.

The JKR approximation works well for high adhesion, large radii of curvature and compliant materials but may underestimate surface forces. An alternative theory have been developed by Derjaguin, Muller, Toporov (DMT) to include noncontact adhesion forces acting in a ring-shaped zone around the contact area [81]. On the other hand, the DMT approximation constrains the tip-sample geometry to remain Hertzian, as if adhesion forces could not deform the surfaces. The DMT model applies to rigid systems with small adhesion and radius of curvature, but may underestimate the contact area. For many SFM's, the actual situation is likely to lie somewhere between these two models [116]. The transition between the models their applicability for SFM problems were analysed elsewhere [120, 143].

In the above discussed theories, purely elastic deformation was assumed. In practice, the SFM tip can cause inelastic deformation or even move portions of the material away from the contact area.

2.3.3
Limitations in Resolution

Similar to all scanning microscopies, the resolution in SFM depends on the effective size of the probe and its modifications which arise from sample-probe interactions. Theoretically, the effective size is determined by the probe geometry and the force-distance dependence between the tip and sample. In addition, the aperture increases because of the tip-sample deformation, surface roughness, capillary forces, and various sources of noise. Experimentally, the resolution is limited by the sensitivity of the force detection system, the image noise, and the scanner precision.

As shown in the previous section, the minimum contact radius may exceed the interatomic distance by far, though the atomic resolution is routinely achieved in many laboratories world wide. Apparently, the resolution in SFM is a more complex problem than just geometric commensurability of the probe size and the surface morphology. At least, one has to treat the problem differently regarding visualisation of the atomic structure and resolution of nanometer sized features.

2.3.3.1
Atomic Resolution

Despite the relatively large contact radius in contact mode SFM (from 1 to 20 nm), an atomic lattice image was obtained for many organic and inorganic substrates, both in the topographic mode and in the lateral force measurement [84, 85, 88, 90]. Usually, the images showed almost perfect order, and lack of the real space structure including defects on the atomic scale. However, there should be no mystery to atomic imaging. One has to clearly distinguish between the real-space resolution and visualisation of the crystalline lattice. Some of the anomalous imaging phenomena on the atomic scale were explained based on Moiré mechanism which can generate periodic fringes within the interaction region [241–243]. The mechanism is illustrated in Fig. 18 by a one-dimensional model of the lateral sliding of two lattices over each other at a given height. The signal on the detector represents a superposition of two lattices k and l as $I(x)=(1+\cos k(x+e)(1+\cos lx)$. The total signal measured by the probe microscope is given by $S=\int_0^{\Delta} I(x)dx$, where Δ is the width of the tip "aperture" over which the signal is measured. Periodic variations of the cantilever deflection upon the multiatom contact was corroborated by computer simulations [244, 245].

It is important to note that the lattice image in Fig. 18a does not depend on the interaction type and even on the aperture size provided the lattices are perfect and the k/l ratio is an integer number. If there are defects in the real space structure and the distance between them is comparable with the contact radius, the phase relation between the lattices can be distorted and the regular structure will vanish (Fig. 18b). This will also occur when the ratio between periodicities of the tip and surface is irrational. In both cases, a small aperture is required to maintain a reasonable level of the measured signal.

Therefore, in most cases the scanning force microscope gives a lattice image similar to diffraction techniques. Visualisation of non-periodic structures or lattice defects, which means the true atomic resolution is exceptional and practically not attainable for polymers [58, 236, 246]. The smallest defects observed by conventional SFM are linear dislocations whose lengths exceed the contact diameter [247–249]. To approach the true atomic resolution, the aperture must be decreased as far as possible by using sharper tips and operating at lowest measurable forces to minimise the contact area. For example, to achieve the contact

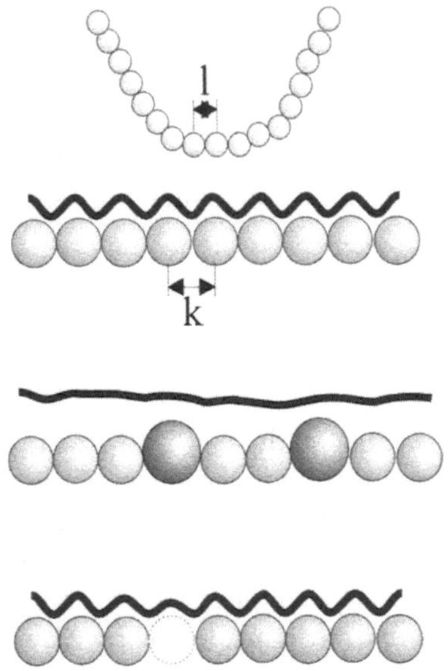

Fig. 18. One-dimensional model for visualisation of the surface lattice by a blunt tip based on the Moiré mechanism. **a** The signal, which is measured by the detector during lateral sliding of the two lattices over each other, represents a superposition of the two wave vectors k and l; **b** The regular pattern vanishes when the correlation between the lattices is disturbed by the irregular surface structure; **c** The regular pattern retains though a single atom defect emerges in the surface lattice

diameter below 5 Å, the favourable experimental conditions would be E= 100 GPa, R=1 nm and a work of adhesion of 100 mJ/m^2.

Any distortion of the sample, in particular plastic deformation, can disturb lattice imaging and makes impossible visualisation of the lattice defects. Operation in liquids results in smaller sample deformation and might even prevent destruction of tiny asperities ant the tip end [250]. By minimising the adhesion forces, Binnig et al has observed different types of point-like defects on the (1014) cleavage surface of calcite [184]. In contradiction to this approach, Magonov et al. demonstrated well-resolved atomic-scale images of inorganic layered crystals which were scanned with high forces up to several hundred nanoNewtons [58]. It has been suggested, that under the strong repulsion condition, a few outermost atoms might dominate in the force towards detection of point-like defects.

In non-contact SFM, the effective contact diameter of a spherical tip (*R*) and a flat surface is given by the area of a flat disk whose interaction force with the surface is the same as that of the tip-surface interaction at the same surface separation D [77]. The disk diameter can be calculated as $d=2\sqrt{RD}$. For a typical set of experimental parameters such as R=10 nm and D=1 nm, the effective interaction diameter is about d=6.3 nm. In practice, the lateral resolution in non-

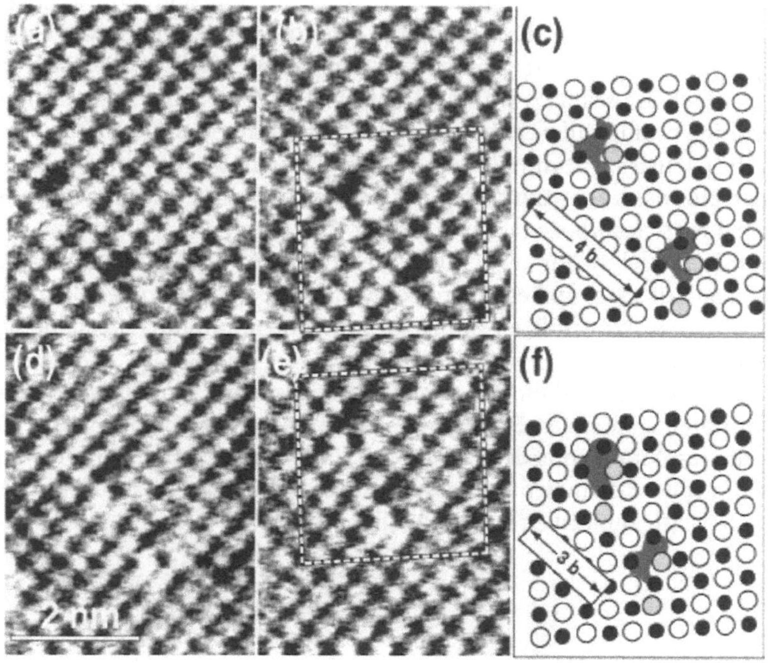

Fig. 19. Series of images (damping signal at constant frequency) of a pair of point defects. **a** and **b** were acquired consequently (10 min time lapses); **c** is a schematic drawing of the observed structure; **d** and **e** were acquired after a sudden change of the distance between the defects by the amount a_0=5.6 Å (lattice constant of NaCl) during 70 and 110 min after (**a**), respectively; **f** illustrates the reduced separation. Courtesy of R. Lüthi [98]

contact mode SFM is rarely better than 5 nm, which is mainly limited by the tip-sample separation. On the other hand, because of the capillary instabilities, the non-contact operation at distances lower than 1–2 nm is hardly possible [242]. Recently, true atomic resolution on Si and NaCl (Fig. 19) was achieved by means of a home-built vacuum SFM which enables stable operation in a near contact regime [98]. The microscope uses stiff cantilevers with conductive tips, operates at resonance conditions (f=153 kHz) and large amplitudes (tens of nanometers). To prevent the jump-to-contact and keep the closest separation between the oscillating tip, a special feedback mechanism based on a tunnelling current between the tip and sample was developed.

2.3.3.2
Nanoscopic Resolution

Unlike the lattice imaging which can be performed by blunt tips, the finite size and 3D shape of the tip become important for imaging of the surface structure on the nanometer scale, where the tip shape and the surface topography may superimpose. Compared to the atomic resolution, different and more simple criteria can be developed for the nanoscopic resolution based on deconvolution of the topographic signal [251]. For example, the sample consists of densely packed spheres which are scanned with a 10 nm tip (Fig. 20a). If the spheres are rigid, the resulting profile will be a smooth curve with a dimple at the middle point between the particles. The dimple depth can be calculated as $\Delta z \approx d^2/(4d+8R)$. If the SFM detector has a sensitivity of ca. 0.1 nm, the dimple of 0.1 nm will be detected for the particles whose radius is greater than 3 nm.

Another model system, which is quite typical for SFM measurements, consists of a pair of sharp spikes separated by distance *d*. The dimple in Fig. 20b can be calculated as $\Delta z = d^2/8R$, e.g. for $\Delta z = 0.1$ nm, the minimum detectable separation will be 2.8 nm.

The above considered models assume that the sample is rigid, which is not the case for polymer samples. For example, flattening of the spheres in Fig. 20d reduces the resolution considerably. A typical deformation of 0.5 nm results in a minimum sphere diameter of about 6.4 nm detectable by a 10 nm tip. A few examples of single molecule resolution will be discussed in Sect. 4.

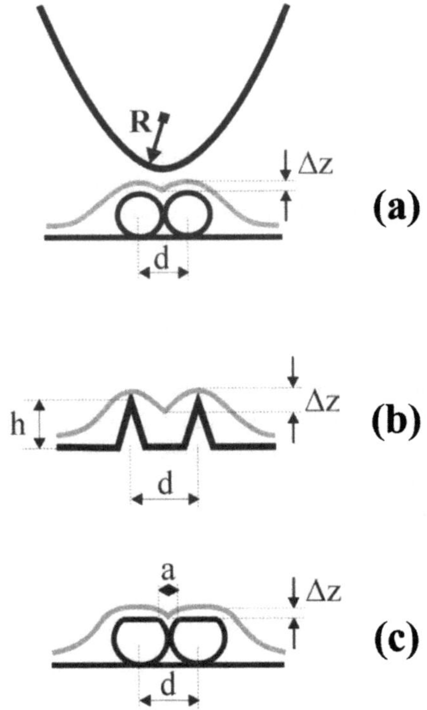

Fig. 20. Different surface morphologies illustrate limitations in the nanoscopic resolution of the scanning force microscope . **a** – two rigid spheres, **b** – two rigid spikes, **c** – two soft spheres. While Δz is the lower limit for the dimple to be resolved by the tip of radius R, d corresponds to the lateral resolution of the SFM tip

To summarise, the nanoscopic resolution is determined by the tip geometry and the surface deformation. Sharper tips and smaller deformations are recommended to resolve three dimensional objects of the nanometer size. However, these requirements are controversial. As the tip radius is reduced, the local pressure on the sample increases inversely to the radius. This dependence arises because the contact area is proportional to the square of the radius, whereas the interaction forces scale roughly with the tip dimension. Therefore, the resolution can be only improved if sharper tips will be used in combination with lower forces.

3
Characterisation and Manipulation Tools for Nanotechnology

3.1
Microphase Separation in Block-Copolymer Films

It is a routine SFM experiment to investigate the heterogeneous structure of polymer blends and composites containing micrometer sized domains [69]. A less trivial problem is to resolve and characterise the features on the nanometer scale (around 10 nm), which are comparable to the tip size and the contact area. Typical systems, which demonstrate microheterogeneous structures, are block copolymers consisting of chemically different and physically incompatible blocks, e.g. A and B. As a result of the interconnectivity of the blocks, block copolymers undergo microphase separation, where the size of the microdomains is restricted to the molecular dimensions. One can distinguish between AB diblock copolymers and triblock copolymers (ABA and ABC).

In the bulk, diblock copolymers show different morphologies consisting of spheres (the cubic phase), rods (the hexagonal phase) and sheets (the lamellar phase), including peculiar transition structures [28, 252, 253]. The morphology type depends on the copolymer composition and a balance between the interfacial energies, the curvature and the chain stretching [254–256]. In the proximity of a solid wall or free surface, the interaction of the blocks with the boundary phase (e.g. air or substrate) favours preferential segregation of one of the blocks at the interface [257, 258]. This can induce parallel orientation of the microdomain structure with respect to the interface plane. For example, symmetric diblock copolymers assemble into a multilayered film whose thickness is an integer number of the equilibrium repeat period [10, 259–261]. The quantitation of the film thickness can be easily documented by SFM using the topographic mode. Figure 21 shows a peculiar surface morphology which was formed upon annealing at 170 °C a thin film of PS-*b*-PMMA. Interfacial interactions caused "dewetting" of the film and resulted in a ziggurat-like structure of alternating PS and PMMA layers. The height of the circular terraces L=29±1 nm corresponds to the equilibrium period of the copolymer.

The cylindrical morphology occurs upon the phase separation of nonsymmetric diblock copolymers as well as ABA triblock copolymers. In thin films, the

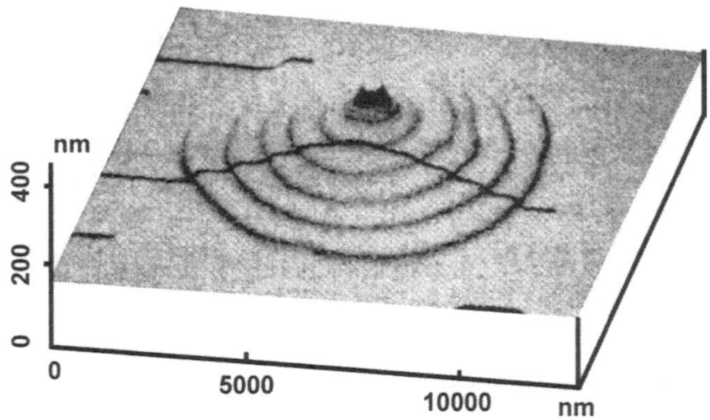

Fig. 21. Ziggurat-like structure obtained by SFM after annealing at 170 °C for one day a thin film of symmetric P(S-*b*-MMA) diblock copolymer (M_w=57,000 g/mol). A step height of 29±1 nm corresponds to the lamellar periodicity. Reproduced from [261]

general trend in minimising of the interfacial energy can cause a transition from cylinders to lamellae [258, 262, 263]. Upon approaching the equilibrium, the lower surface-energy block concentrates at the surface and covers the cylindrical morphology by a thin overlayer. Indeed, a featureless structure of a styrene-butadiene-styrene (SBS) copolymer was observed by SFM under light tapping conditions [264–267]. However, as the tip-sample force was increased, both the height and phase images gradually underwent local changes and eventually revealed the cylindrical pattern underneath the topmost butadiene rich layer. This observation is in contradiction to the earlier studies showing that the cylinders tend to reorient with their main axis perpendicular to the surface [268]. Presumably, the perpendicular orientation was caused by quenching a non-equilibrium structure upon the fast solvent evaporation. The difference in the cylinder orientation was observed upon variation of the film thickness with respect to the cylinder spacing [269]. While thick SBS films (Kraton D-1102 C) displayed only the parallel orientation, both orientations were observed in the ultrathin films of a non-integer thickness (e.g. 49 nm). To summarise, with SFM alone, it is very problematic to distinguish between the cylindrical and edge-on lamellae morphology as well as between the covered and naked microdomain structure. Complementary techniques must be used to learn more about the true topography and chemical composition of the surface, and orientation of the domains near the surface.

Recently, significant advances have been made in controlling the microdomain orientation in diblock copolymer films [10, 11, 270–284]. When the copolymer is placed between two solid interfaces, the morphology responds strongly to the deviation of the film thickness from the integer number of the layers [271–274]. By adjusting the chemical structure of the interfaces so that both blocks interact with them equally, perpendicular orientation of the microdomains was achieved [270]. Also packing incommensurability between a crystallizable block and an amorphous block can lead to frustration in the confined block copolymer and cause perpendicular orientation of the blocks [275]. Figure 22 shows a thin film morphology of a low molecular weight copolymer of hydrogenated polybutadiene and polyethylene oxide. It was crystallised at 38 °C closed to a triple line caused by the dewetting process so that both PB and PEO block became exposed to the surface.

In ultrathin films (thinner than the equilibrium period) the microphase separation is controlled by the surface interactions following one of the scenarios in Fig. 23. In the strong segregation limit, where one of the blocks interacts strongly with the substrate, lateral phase separation might occur similar to those in Fig. 23b. Upon adsorption of diblock copolymer P(S-*b*-2VP) from a non-selective solvent on mica, P2VP adsorbs preferentially to mica and anchors the PS block to the surface [276–279]. In the thus formed "pinned micelles" [280], protruding islands correspond to PS clusters surrounded by a P2VP film (Fig. 24). The height of the clusters, their lateral diameter, and average spacing depends on the block

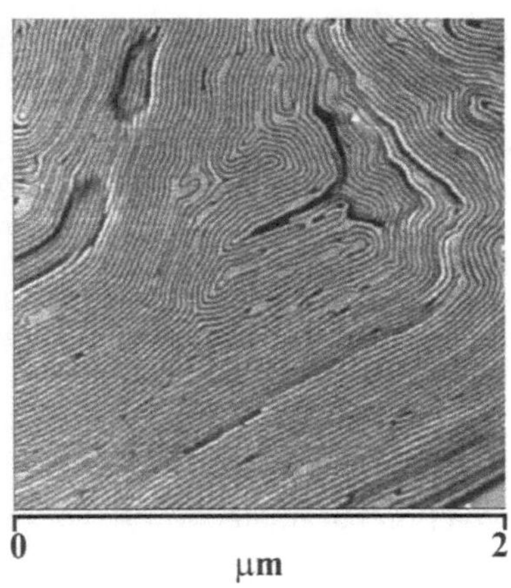

Fig. 22. Phase image of a thin film of a hydrogenated polybutadiene-polyethyleneoxide blockcopolymer (3700–2900 g/mol) on mica. The film was crystallised at 38 °C close to a triple line caused by dewetting. Courtesy of G. Reiter [275]

Fig. 23. Schematic illustration of wetting geometries expected for ultra-thin films of diblock copolymers: **a** – parallel lamellae, **b** – surface (pinned) micelles, **c** – perpendicular lamellae. L corresponds to the equilibrium period of the lamellar morphology

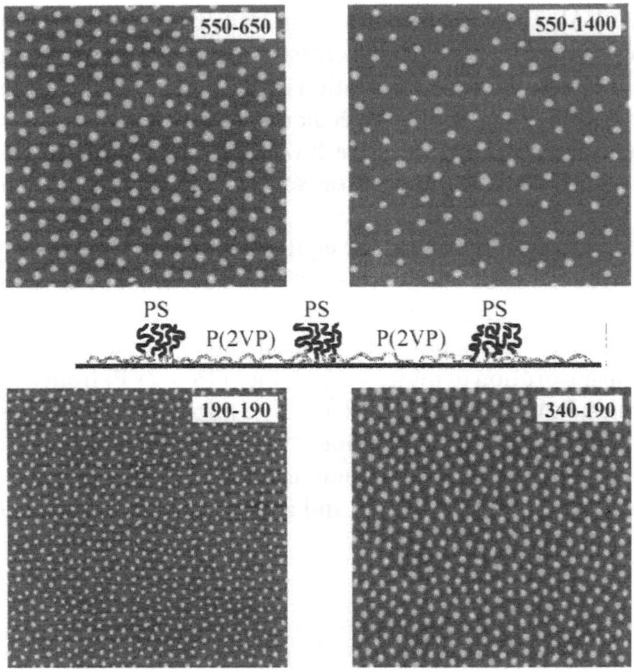

Fig. 24. Regular patterns were formed by adsorption of P(S-*b*-2VP) block copolymer on mica and imaged by contact mode SFM. The micrograph size is 2.5×2.5 μm^2. Depending on the polymer composition, periodicity of the PS clusters varied from 80 to 350 nm and their size changed from 50 to 140 nm [277]. The compositions are indicated above the micrographs and depict the degree polymerisation of PS and P2VP blocks, respectively. The drawing illustrates the pattern origin due to microphase segregation of the polymer blocks on the selective substrate. Courtesy of P. Eibeck and M. Möller

length ratio and the interaction parameters [279]. For example, unlike the globular micelles of P(S-*b*-2VP), worm-like aggregates are formed upon adsorption of P(S-*b*-4VP) resulting in a striped surface pattern as shown in Fig. 25.

A few attempts have been made to generate a surface pattern by adsorption of a diblock copolymer to a chemically heterogeneous surface [281–283]. Figure 26 shows two SFM images of symmetric P(S-*b*-MMA) block copolymers differing in the molecular weight after they were cast on a heterogeneous surface containing Au/SiO_2 stripes with a 60 nm periodicity, respectively. The molecular weight was adjusted with respect to the periodicity of the substrate in order to control microdomain orientation over long ranges. An electric field was shown to be an alternative means of controlling the copolymer orientation [284]. This type of laterally structured films offer many interesting applications in nanoscience and technology [11, 20, 23].

In a selective solvent, diblock copolymers undergo phase separation as well. They form micelles where the core consists of the insoluble block and the shell

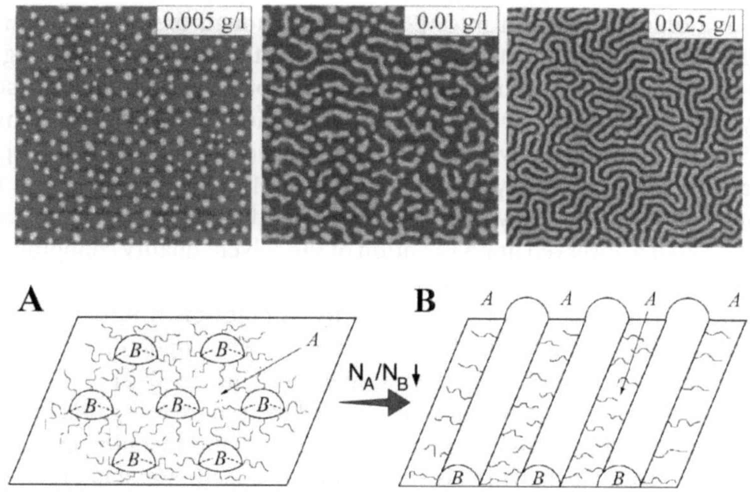

Fig. 25. Surface structures formed on mica upon adsorption of P(S-*b*-4VP) 330–130 from chloroform [279]. The 2.5×2.5 μm^2 height images were obtained by contact mode SFM. Depending on the solution concentration, the morphology developed from small islands on the *left* to lamellar morphology on the *right*. The solution concentrations are indicated above the micrographs. The drawing *below* illustrates the transition from the **A** dot-like to **B** stripe-like morphology caused by variation of the block-length ratio. Courtesy of P. Eibeck and M. Möller

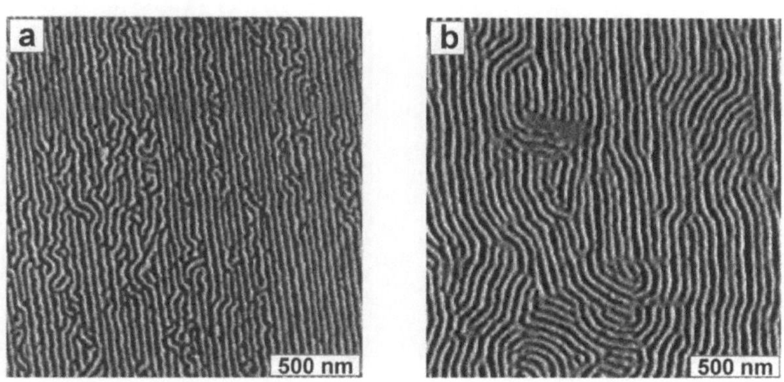

Fig. 26. 50:50 block copolymer of styrene and methacrylate cast from 0.1 wt % solution in toluene onto a chemically heterogeneous surface with Au/SiO_2 stripes with 60 nm. Periodicity and the degree of order of the striped pattern revealed strong dependence on the molecular weight: **a** 113,000 g/mol and **b** 177,000 g/mol. Courtesy of T.P. Russell

is formed of the soluble block [285–289]. Adsorption of the micelles onto a solid substrate, assisted by rapid solvent evaporation, prevents the micellar structure from decomposition and can lead to a laterally ordered film. Spherically shaped micelles of P(S-*b*-2VP) formed a hexagonally ordered structure (Fig. 27a) upon adsorption onto a surface of mica [289, 290]. Although the micellar film is far from the melt equilibrium state, it could remain stable for infinitely long time. When the soluble block becomes considerably shorter compared to the soluble one, cylindrical micelles emerge due to coagulation of the spherical micelles [291]. Similar to blockcopolymer associates in aqueous solution [33, 286], giant micellar threads were clearly documented by SFM in combination with transmission electron microscopy (TEM). A transition from spherical to cylindrical micelles was also observed upon variation of the solvent quality changing the ratio between the physical dimensions of the blocks [288]. Also ABA-type triblock copolymers form micelles in a solvent selective for block A. Micellar structure of thin films of polystyrene-oligothiophene-polystyrene copolymer were identified by tapping mode SFM using the phase imaging mode [292].

The diblock copolymer micelles can be used to prepare inorganic nanoparticles ranging from 1 to 10 nm in diameter [12, 30]. In the so-called microreactor approach, the core block, which is shielded by the shell interacts preferentially with the inorganic molecules in solution causing their precipitation inside the micelles. If the inorganic component was a metal salt, the latter could be reduced to a free metal state by either chemical or temperature treatment. The size of thus obtained metal cluster was controlled by the micelle size depending on the

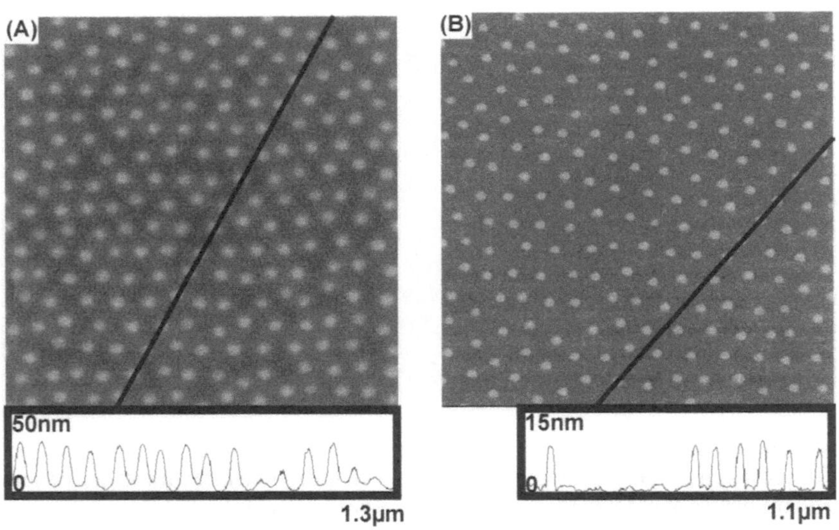

Fig. 27.a – In solution formed micelles of P(S-*b*-2VP) 1700–450 adsorbed from 0.1 g/l toluene solution. The micelles are loaded with tetrachloro-auric acid at a ratio $HAuCl_4$ /2VP=0.5. **b** – The regular pattern of naked gold particles were obtained by oxygen plasma treatment of the micellar film in (**a**). The plasma treatment results in reduction of the gold salt and depolymerisation of the polymer shell. The height images were obtained by contact mode SFM. Cross sectional profiles were drawn along the lines in the micrographs and show the height variation of the micellar and gold films. Courtesy of S. Mößmer, J.P. Spatz, and M. Möller

aggregation number and the degree of polymerisation. Recently, another use of the diblock copolymer micelles has been realised to prepare metal dots by plasma etching of the polymer clad [293,294]. Figure 26b shows an SFM micrograph of a hexagonally ordered array of 5 nm sized gold clusters on a silicon wafer. Both the cluster size and the interparticle distance was controlled by the micellar morphology.

The concept of using block copolymers for preparation of nanoscopically structured material and surfaces was advanced further by introducing a third block in the chain structure [29]. One of the consequences of the multiphilicity and versatility of the ABC triblock copolymers is their tremendous richness and diversity in morphology. One of the most peculiar structures is shown in Fig. 28 where the helices of a polybutadiene microphase are wound around columns of polystyrene which are embedded in a matrix of polymethylmethacrylate. Complementary to the TEM studies of the bulk morphology (Fig. 28a,b), SFM has been used to image the surface structure of the triblock copolymer films. Figure 28c shows the wrapped PS cylinders oriented parallel to the surface, where one

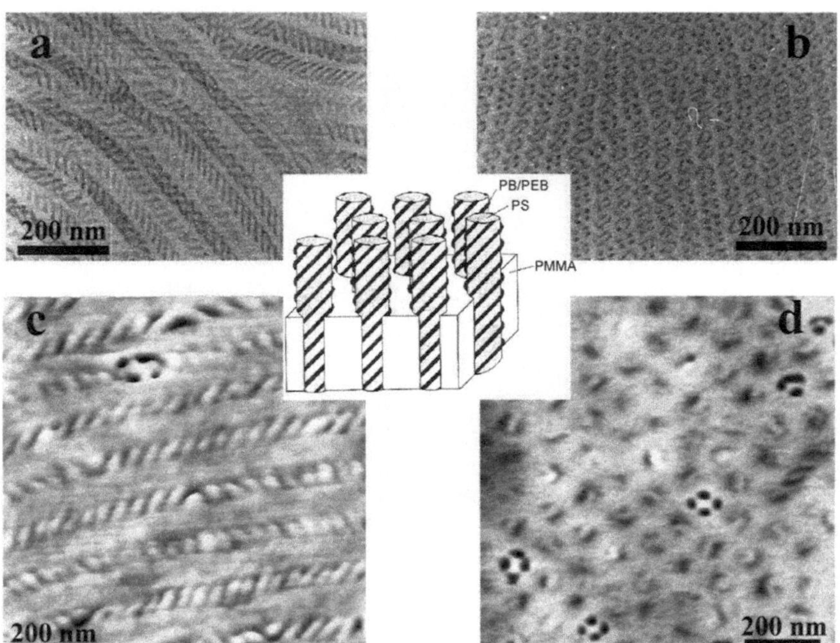

Fig. 28. a, b TEM micrographs of a triblock copolymer p(styrene-butadiene-methacrylate) where PS cylinders are wrapped by the PB block and embedded in the PMMA matrix (courtesy of R. Stadler). Height (**c**) and phase (**d**) micrographs (tapping mode SFM) of the surface morphology of a thick triblock-copolymer film. The phase image shows a 4-point pattern corresponding to the cylinders which are oriented perpendicular to the film surface. Courtesy of A. Mourran and L. Leibler

can clearly see the helical coat of the PB block similar to the TEM micrographs. Eventually, also perpendicular orientation of the columns was observed resulting in a surface morphology shown in Fig. 28d, where the circular cross section of the PS phase is surrounded by a quatroplex of PB domains. Samples with a smaller PB block led to another morphology where PB spheres are located at the lamellar PS/PMMA interface [295]. Apart from the discussed examples, the surface arrangement of the blocks in a triblock-copolymer film is less apparent compared to their diblock analogues. Usually, SFM cannot distinguish between the domains and make clear assignment of the domains to a certain block. Additional means, such as swelling in a selective solvent, can be used to facilitate the interpretation of the surface structure [296].

Also the ultrathin film morphology of triblock copolymers appeared to be very peculiar. Figure 29 shows a microphase separated structure of P(St-*b*-2VP-*b*-MMA) adsorbed from a nonselective solvent on mica. All three blocks were resolved by lateral force measurements. Compared to the diblock copolymer morphologies in Figs. 24 and 25, this represents an exceptional example where three chemically different components could be arranged in a regular pattern with a long range coherence.

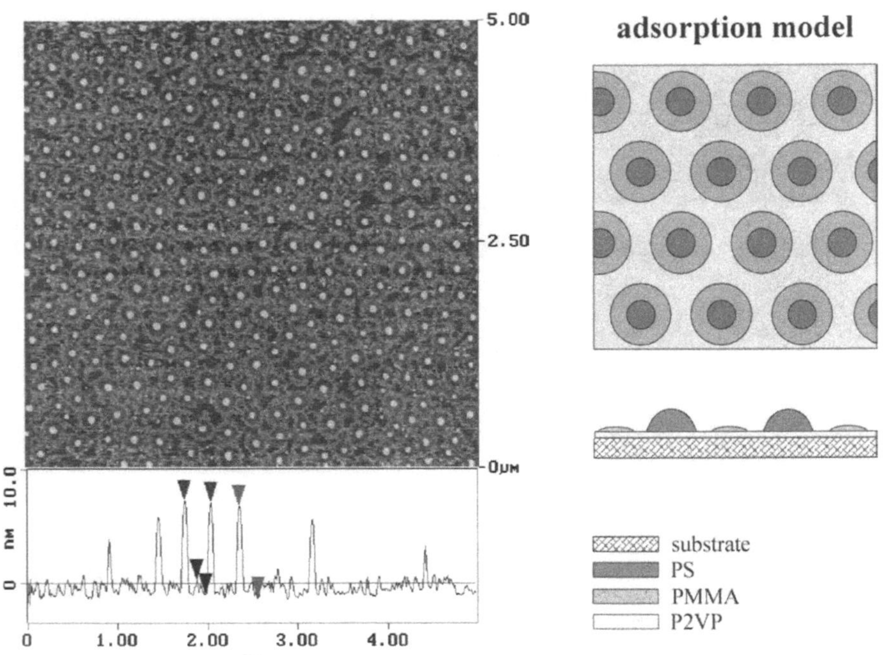

Fig. 29. Ultrathin film morphology of triblock copolymer P(S_{800}-*b*-2VP$_{430}$-*b*-MMA$_{730}$) adsorbed from a dilute solution in chloroform onto mica as observed by contact mode SFM. The drawing on the *right hand side* shows the location of the blocks on the substrate. Courtesy of P. Eibeck and M. Möller

3.2
Surface Wetting on the Nanoscale

Wetting and spreading phenomena control many life and technological processes such as breathing, lubrication, soldering, painting, gluing and many others. Thermodynamics and kinetics of wetting are rather well understood macroscopically for simple liquids [297–299]. The static contact angle and the equilibrium film thickness were explained by the interplay between the interfacial tensions and the surface pressure caused by long range forces, whereas hydrodynamic theories were employed to describe the spreading kinetics. Figure 30 depicts typical wetting morphologies observed experimentally and predicted theoretically in thin liquid films.

Due to the chain architecture and the large size of the macromolecules, the wetting behaviour of polymer liquids can be different from that of simple liquids. The effect becomes particularly strong when the dimension of the liquid phase, e.g. film thickness and droplet diameter, approaches the dimension of the polymer coil. In addition to the spreading coefficient and the surface pressure effects, entropic elasticity of the polymer chain provides a strong contribution to the free energy for a constant volume $V_0=Ad$:

$$\Delta F=F_{dry}-F_{wet}=[S+P(d)-E(d)]A,$$

where $S=\gamma_s-\gamma_l-\gamma_{sl}$ is the spreading coefficient, $P(d)=-H/12\pi d^2$ – surface pressure of the long range van der Waals interaction (H – Hamaker constant, d – film thickness), and $E(d)=3/2kTn[Nl^2/d^2-1]$ is the decrease in the configurational entropy due to compression of a Gaussian coil toward the surface, where $n=d/Nv$ – number of polymer molecules per unit area [300].

The entropic arguments have been used to explain many unusual phenomena, such as dewetting of polymer brushes by the melt of the same polymer [301] and enrichment of the surface with lower molecular weight component [302], which are not observed for simple liquids. Also wetting dynamics is strongly af-

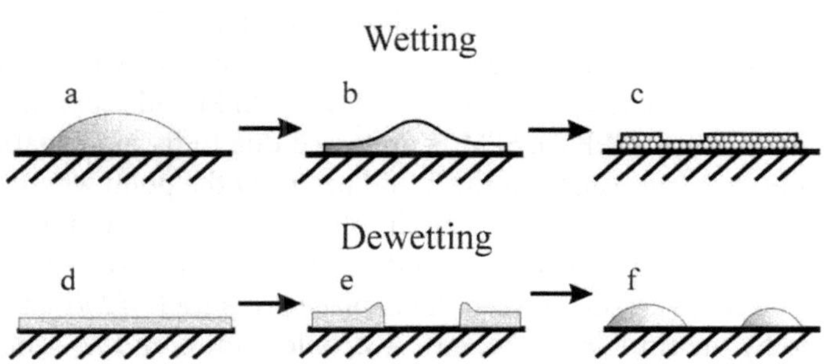

Fig. 30. Different types of the momentary morphologies which are typically observed during wetting (**a–c**) and dewetting (**d–f**) events on a solid flat substrate. **a** droplet, **b** spherical cap with a precursor film, **c** thin film (eventually with a multilayer structure), **d** thin liquid film, **e** ruptured film with rims at the dewetting front, **f** droplets

fected by the chain structure. Anomalous glass transition and stepwise viscosity-versus-thickness dependence were observed in the vicinity to the interface [303, 304]. Depending on the molecular weight, the mechanism of spreading represents a combination of chain slippage and shear flow [305–308].

The outlined problems correspond to the most typical system where three phases, e.g. liquid, gas and solid, are brought in contact. Additional wetting geometries can occur when the liquid phase consists of two subphases, e.g. mixture of incompatible polymer liquids, and/or the substrate surface exhibits variations in chemical composition. In these cases, the interfacial interactions will strongly interfere with the phase separation inside the film. Laterally ordered polymer films might be formed due to the preferential wetting of the patterned substrate by one of the liquid phases.

Therefore, the interfacial behaviour of liquids contains a great set of unsolved problems, which are particularly intriguing for complex polymer fluids. It is an experimental challenge to measure nanoscopic properties such as line tension and the microscopic contact angle as well as to follow propagation of the precursor film and ordering of the microheterogeneous systems. The progress in this field rely on SFM techniques which showed strong advances in probing the surface structure of liquids. In this section we will discuss a few examples where SFM provides a unique information on the wetting properties.

3.2.1
Apparent Contact Angle and the Shape of the Contact Line on the Nanoscale

The contact angle is used to characterise the wetting properties of the surface with respect to a certain liquid [6]. Theoretically, the equilibrium contact angle can be calculated from the Young's equation. In practice, one measures an apparent contact angle which depends on local variations in the chemical composition and the surface roughness. The surface heterogeneity results in the contact angle hysteresis and variations of its value over the surface. On the nanometer scale, the droplet profile and the wetting ability are affected by the long-range surface forces [309, 310]. Also the effect of the line tension should be taken into account when the droplet curvature approaches the nanometer scale [311, 312].

In-situ imaging of microscopic droplets became possible with the introduction of scanning probe techniques such as SNOM, SFM and scanning polarisation force microscopy (SPFM). SFM is preferred due to its high spatial resolution, enabling characterisation of the liquid profile at the point where the liquid meets the surface. It is essential to scan droplets with a minimum possible force F to minimise perturbations of the liquid surface. The perturbations caused by the tip depend on the ratio between the Laplace pressure $P_L=2\gamma/r$ and the pressure exerted by the tip $P_T=F/A$, where r – droplet radius and A – contact area. Small droplets and lower forces promote non-destructive imaging. The first microscopic images of liquids were obtained by SPFM, where the tip was kept far away (≥10 nm) from the surface [99, 169, 313]. Long range electrostatic forces were employed to perform non-contact imaging of thin layers of water on mica as well as tiny droplets of KOH water on graphite. The contact angle appeared to be somewhat lower than the macroscopic contact angle, and furthermore, it decreased with the droplet size. This behaviour was explained by the surface pressure and line tension effects [313].

The non-contact conditions result in degradation of the spatial resolution (Sect. 2.3.3). Although the vertical resolution remained excellent (0.1 nm), the lateral resolution of SPFM was of the order of the tip-sample distance, i.e. 20–40 nm. Recently, also SFM have been used for imaging microscopic droplets [314–316]. For example, droplets of low-viscosity liquids such as aqueous solutions of $CaCl_2$ equilibrated with an ambient humidity of ca. 50% [316] were imaged by tapping mode SFM. The changes in the cantilever response during the droplet imaging were ascribed to the attractive van der Waals forces and capillary bridging between the tip and sample. The nanometer resolution enabled visualisation of tiny undulations of the contact line on a heterogeneous substrate. From the modified Young equation the line tension on hydrophobic areas was calculated to be about 5×10^{-10} N, and appeared to be in general agreement with the theory [317, 318].

Also complex liquids such as glycerine [314] and carbosilane dendrimers [315] could be imaged with tapping mode SFM. For the dendrimers, the topographic and phase contrast have been ascribed to a relaxation transition of the liquids into a viscoelastic state at a tapping frequency of ca. 300 kHz. Figure 31 shows three dimensional images of the dendrimer droplets captured on different stages of their spreading on mica. A spherical cap in Fig. 31a transforms slowly into a pancake structure with a reservoir in the central part (Fig. 31b). The droplets spread via development of a thin precursor film which extends ahead the macroscopic part as shown in Fig. 30b. The film thickness was measure to be about 1.6 nm which corresponds to the size of a dendrimer molecule.

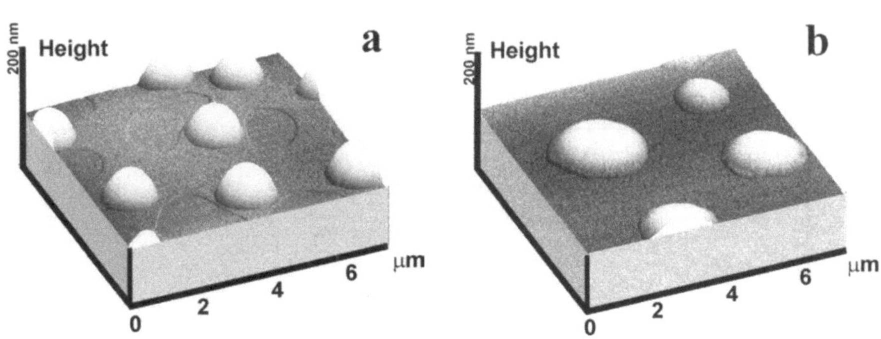

Fig. 31. Three-dimensional images of droplets of the carbosilane dendrimer with hydroxyl end groups obtained during spreading at ambient conditions on a mica substrate [320]. The images corresponds to different duration of spreading (**a** – 10 min and **b** – 24 h), where a precursor film of 2 nm in thickness develops at the contact line in (**b**). The thickness corresponds to the diameter of a dendrimer molecule

The dendrimer droplets were used as a tool to examine surface heterogeneity on the nanoscale [319]. Figure 32 shows two pairs of SFM images of the droplets deposited on mica (a,b) and a copolymer film (c,d), respectively. On mica, the hydroxyl terminated dendrimer showed the autophobic or pseudo-partial wetting behaviour as the droplets sit on a thin film of the same material [312, 320].

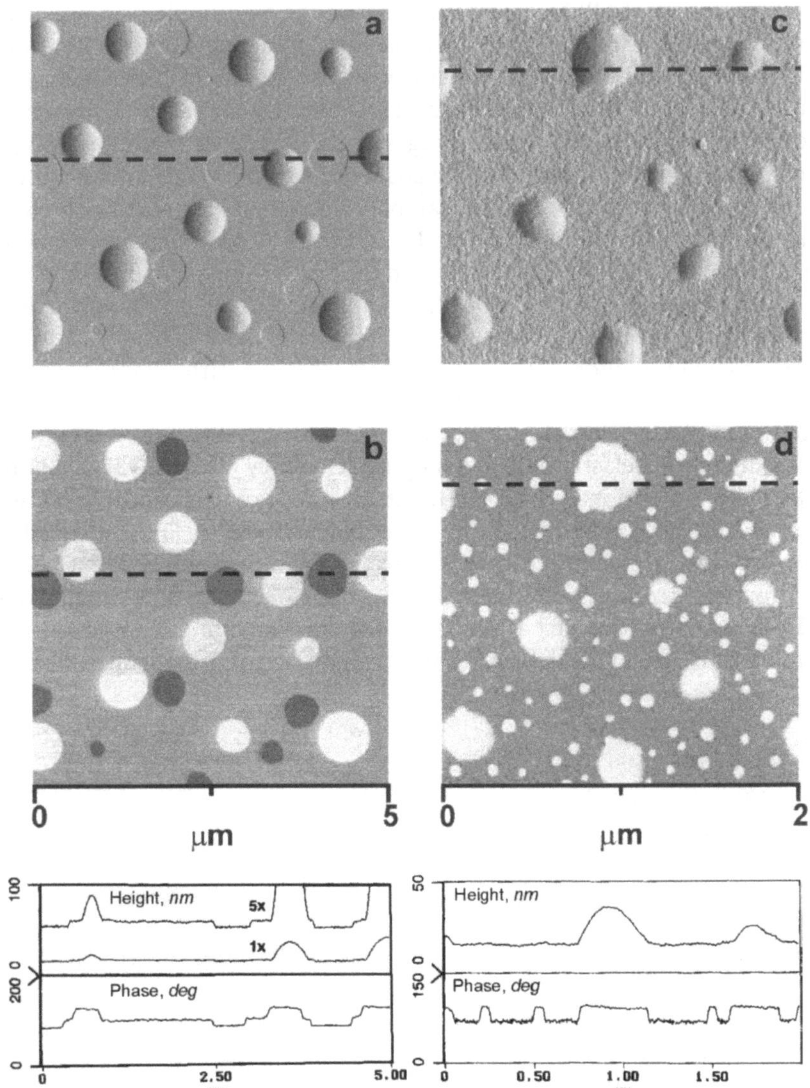

Fig. 32. Amplitude (**a**,**c**) and phase (**b**,**d**) SFM micrographs demonstrate autophobic wetting of mica (**a**,**b**) and semifluorinated copolymer (**c**,**d**) by carbosilane dendrimer with hydroxyl end groups. Fluid droplets with a contact angle of about 8.7 degrees in (**a**) and 18.5 degrees in (**c**) were determined after 24 h equilibration at room temperature from cross sectional profiles recorded along the reference lines indicated on the (**a**) and (**b**) respectively. Reproduced from [319]

The contact angle could be determined directly at the contact line or calculated from the spherical shape of the droplet to be 8.4 ° and 8.9 °, respectively. On the heterogeneous copolymer surface, the droplets were not perfectly spherical and the contact line exhibited deviations from the circular contour. The dendrimer droplets (γ=29.3 mN/m) displayed a greater contact angle of ca. 18.5 ° on the copolymer film (γ=23 mN/m) compared to the droplets on mica. Deformation of the droplets by the tapping tip and thus caused underestimation of the contact angle have been discussed in [319].

3.2.2
Dewetting of Thin Polymer Films

Modern coating technologies require increasingly thinner polymer films. This requirement is opposed by the surface pressure and the chain elasticity. Below a certain equilibrium thickness, the film is either metastable or even unstable and tends to break into droplets regardless of the chemical structure of the substrate [321, 322]. Anomalous wetting behaviour was observed for amphiphilic polymer films whose stability is controlled by the orientation of the surface active moieties [323, 324]. All these phenomena belong to the dewetting problem.

Scanning force microscopy became widely used for monitoring of the dewetting process [323–329]. Complementary to optical microscopy, SFM provides nanoscopic information about the topographic profile of the dewetting liquid as well as the microstructure of the dewetted surface. Due to the high resolution, SFM is able to capture evolution of the dewetting morphology at much earlier stages than optical microscopy. Nucleation of nanometer sized holes and droplets can be readily visualised by SFM. Figure 33 shows a series of images obtain at different stages of the dewetting of a smectic polymer film [323]. The dewetting process starts from the film rupture and formation of small holes (a). The holes grow until a certain critical size (b) followed by formation of towers (c). Finally, the towers, which might grow via a layer penetration mechanism [330], collapse into droplets (d).

Besides the autophobic behaviour promoted by the smectic structure of the semifluorinated polymer films, the chain elasticity alone may cause the film rupture as well [324]. Figure 34 shows two SFM micrographs obtained on a microphase separated film and a homogeneous polymer film of polymethacrylates differing in the length of the hexafluoropentane side chains ($HFPO_5$ and $HFPO_3$, respectively). Although both polymers were in a fluid state, ca. 50 K above the glass transition temperature, their wetting behaviour was qualitatively different. While the phase separated $HFPO_5$ polymer showed autophobic droplets sitting on an oriented monolayer of itself, the homogeneous $HFPO_3$ polymer resulted in a uniformly thick film whose thickness is controlled by the chain dimensions.

The spreading coefficient, and therefore, wetting properties of the substrate can be varied by chemical modification of the substrate surface or the liquid itself. From SFM observations, the wetting behaviour of PS on SiO_x substrates was changed by partial sulfonation [331]. Unlike the unmodified PS, which readily

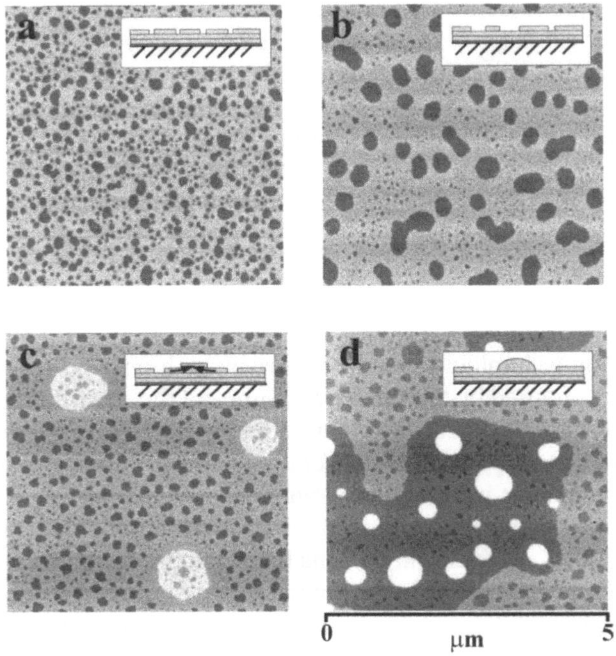

Fig. 33. Self-dewetting of a polymer smectic film into droplets was monitored by tapping mode SFM. The micrographs were measured at room temperature after annealing of a 10 nm thick film at 115°C during different time intervals: **a** – 0 min, **b** – 15 min, **c** – 2 h, **d** – 4 h. The annealing temperature was 20 °C above the bulk isotropisation temperature. The polymer consists of a polymethylmethacrylate backbone with perfluorooctyl side chains [323]

dewets the substrate, a strong inhibition of dewetting occurs already for low sulfonation of ca. 2.3 mol %.

3.2.3
Autophobicity of Polymer Networks and Brushes

The wetting properties of immobilised polymer layers such as networks and brushes play an important role in their technological applications. Their wettability is accompanied by penetration of the polymer melt into the underlying layer [301, 332–337]. If the melt chains were relatively short they invaded the grafted layer and complete wetting occurred. On the other hand, interpenetration of the long polymer chains was inhibited by the stretching of the brush chains. The loss in the configurational entropy leads to a negative spreading coefficient $S \cong -3/8kT/\lambda a$ for a penetration depth $\lambda \cong (12/\pi^4)^{1/3}(Na/\sigma)^{1/3}$, where σ^{-1} – grafting density, and a – monomer (Kuhn) length. Therefore, the entropic penalty limits the penetration and results in partial dewetting though the polymer melt and the layer are of the same chemical nature [333]. This phenomenon is similar to "wetting autophobicity" observed by Zisman for small mole-

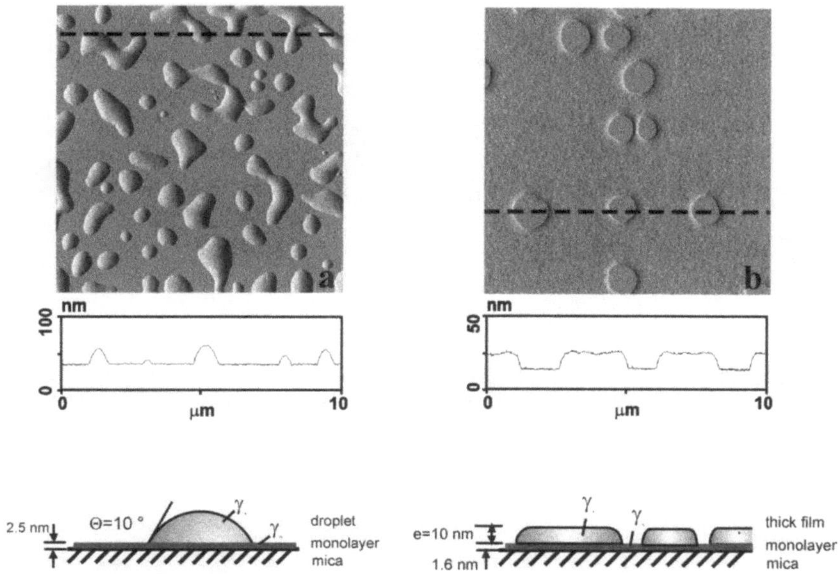

Fig. 34. SFM amplitude images of the oligo(hexafluoropropene) substituted PMA films on mica after annealing at room temperature for 50 h [324]. Depending on the number of HFPO units, polymers exhibit either **a** autophobic wetting behaviour for $HFPO_5$ or **b** stable coverage with a 10 nm thick film for $HFPO_3$

cules, where the droplet formation was discussed in terms of dispersion energy [320].

Figure 35 shows droplet of PS (91 K) melt which was formed on a PS (91 K) brush layer. For this system, small contact angles are expected as estimated from $\Theta \cong 2(|S|/\gamma)^{1/2}$. From SFM, the contact angle as low as 3 ° appeared to be lower that the calculated value of 15°. In addition to the entropy considerations, the effect of grafting or cross-linking induced roughness on the spreading coefficient was taken into account for polymer networks [336, 337].

3.2.4
Propagation of the Substrate Roughness

One common problem in preparing uniform and smooth films is the propagation of the substrate roughness to the film surface. For simple liquids, the substrate roughness became undetectable for films thicker than ca. 10 nm due to the smoothing tendency driven by surface tension [338, 339]. Recently, the propagation of roughness from a laterally structured Si substrate by polymer films was investigated by SFM [340]. The variation of the surface roughness was measured as a function of the film thickness and the interfacial energy. Figure 36 shows

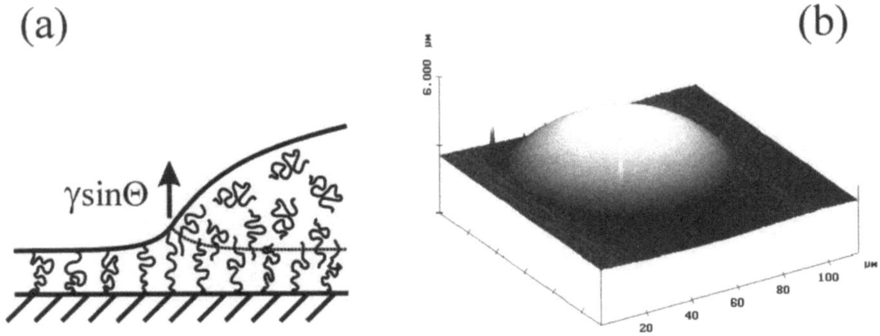

Fig. 35. Equilibrium profile of a polystyrene (M_w=91 K) droplet on a polystyrene brush (M_w=91 K) surface of a PMMA/PS film as obtained by tapping mode SFM [301]. The dewetting occurred after annealing of the PS film at 150 °C during 20 h. Courtesy of A. Mourran and L. Leibler

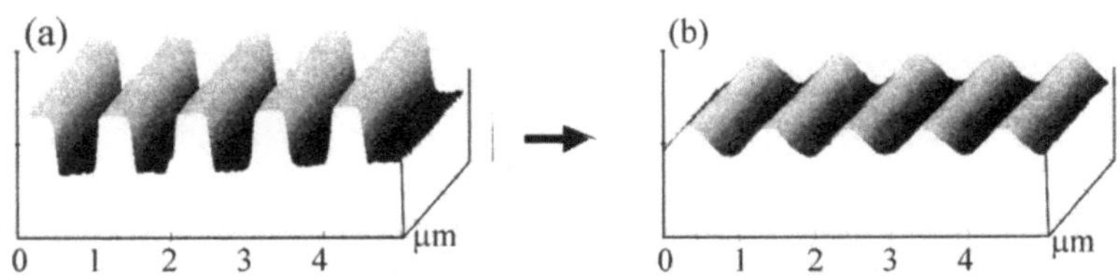

Fig. 36. SFM images obtained from **a** the bare trapezoidal grating of Si with h=17 nm and **b** the same Si grating covered with a PMMA (M_w=330 K) film, 75 nm thick, which was annealed for 72 h. Reproduced from [340]

two SFM images obtained from a bare Si grating and the same grating covered by a PMMA (330 K) 75 nm thick film. From the topographic images one can see that the surface roughness decreased from 17 nm to 11.5 nm after annealing at 170 °C for 72 h. Unlike simple liquids, the roughness became completely attenuated only after the film thickness increased four times the height of the grating. The damping tendency was shown to be promoted by lower interaction of the film with the substrate.

3.2.5
Phase Separation in Thin Polymer-Blend Films

Compared to bulk polymer mixtures, the interfacial behaviour of polymer blends is essentially different [341]. The demixing process in thin films is strongly affected by the thin film confinement and the interfacial interactions of the blend components with the confining phases (e.g., substrate and air). Even in the one-phase region of the phase diagram, preferential segregation of the components at one of the interfaces leads to a certain composition profile as a function of the distance from the free surfaces and the substrate plane [342, 343]. In the unstable two-phase region, the composition profile in the z-direction can be represented by superposition of two spinodal waves, originating from the free surface and the substrate respectively [344–347]. Depending on the ratio between the spinodal wavelength in the bulk and the given film thickness, different phase separation morphologies can be momentary observed as the system drives to its equilibrium structure [348–356]. Figure 37 shows some of the film morphologies of polymer blends which can be predicted from the interfacial energies relation [357].

The following problems have been studied experimentally using different SFM techniques: (i) the surface topography and its dependence on the preparation conditions, (ii) the chemical composition of the surface, and (iii) the role of the substrate structure in polymer demixing. Here, one has to distinguish between asymmetric blends, where one of the components concentrates at the surface and the other at the substrate interface, and symmetric blends where one of the components segregate at both interfaces.

The bilayer morphology of thin asymmetric films of may be unstable. A regularly corrugated surface structure of the films was ascribed to spinodal transition into a laterally phase separated structure, where the surface morphology depended on the polymer incompatibility and the interfacial interactions [347, 348]. Recently, the phase separation and dewetting of thin films of a weakly incompatible blend of deuterated PS and poly(p-methylstyrene) have been monitored by SFM [349, 350]. Starting from a bilayer structure, after 454 h at T= 154 °C the film came to the final dewetting state where mesoscopic drops of

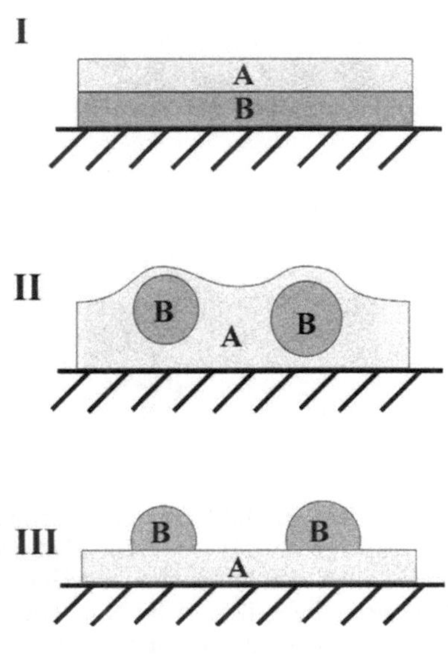

Fig. 37. Theoretical predictions for thin film morphology of a phase-separating polymer mixture. **Case I:** $\gamma_A<\gamma_B$ and $\gamma_{AB}<<\gamma_B-\gamma_A$, where γ_A and γ_B are the surface tensions of components A and B, respectively, and γ_{AB} is the interfacial tension. **Case II:** $\gamma_A<\gamma_B$ and $\gamma_{AB}<<\gamma_B-\gamma_A$. **Case III:** $\gamma_A<\gamma_B$ and $\gamma_{AB}>>\gamma_B-\gamma_A$. Drawn after [357]

PpMS sit on a homogeneous layer of dPS. However, if the surface interactions are sufficiently strong to overwhelm thermal fluctuations, the layered domains structure can be stable for a long time [351, 352]. On a SiO_x surface, the surface morphology of a PS/PMMA blend was ascribed to the layered demixing where the lower-surface-energy PS phase segregates preferentially at the polymer/air interface (Fig. 38a–c). The morphology exhibited strong dependence on the substrate type. A complete phase inversion occurred the alkane monolayer was used as a substrate. Even though PS has a lower surface energy than PMMA, it almost completely excluded from the air surface (Fig. 38d–f).

Symmetrically segregating polymer blends have also been studied recently [353–356]. In contrast to the bilayer structure due to the asymmetric surface segregation, the thin film confinement of the symmetrically segregating phases can suppress the development of surface directed spinodal decomposition [345, 353]. SFM was used to examine the surface morphology of dPS/PVME films where the PVME reach phase tends to segregate at both interfaces with regards [355]. Topographic images displayed a corrugated morphology which was ascribed to dPS domains encapsulated by a thin layer of PVME. The domains were identified due

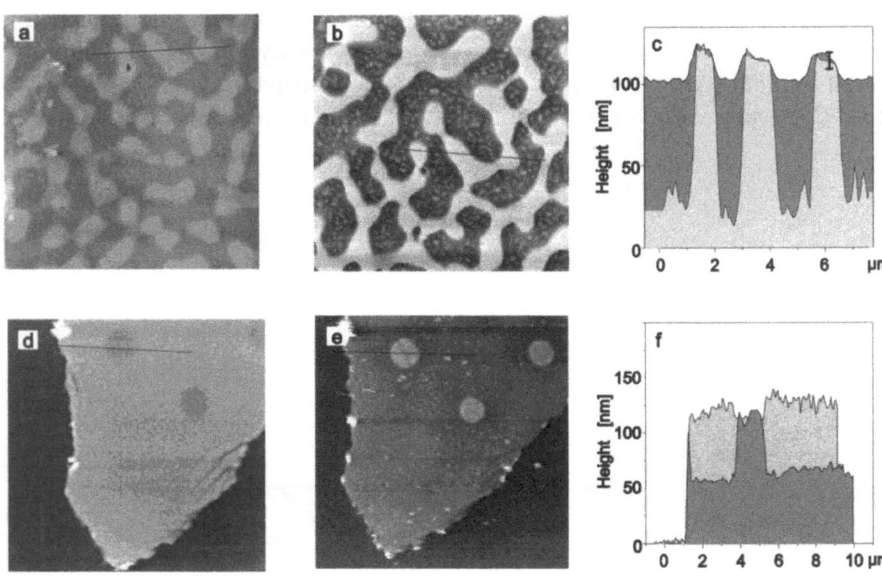

Fig. 38. Substrate dependence of the PS/PMMA domain structure spin-cast from THF: **a-c** on SiO_x; **d-f** on octadecyl mercaptane (ODM) [352]. The SFM images have lateral dimensions of 14×14 μm^2. **a,d** as spin cast; **b** after immersion in cyclohexane to remove PS. **e** after immersion in acetic acid to remove the PMMA-rich phase. The cross-sections (**c,f**), which were recorded along the lines in (**a,b,c,d**), reveal the vertical distribution of the PS (*dark grey*) and PMMA (*light grey*) phases. The error bar in (**c**) indicates the accuracy of the superposition procedure. PMMA preferentially adsorbs on the more polar SiO_x surface to form a homogeneous layer next to the substrate. On the ODM a PS/PMMA bilayer is observed. Courtesy of U. Steiner

to the strong mechanical contrast between relatively hard dPS- and soft PVME domains. Reverse temperature transition of the phase-separated blend films into the one-phase region resulted in dissolution of the undulating surface structure [356]. This confirms the phase separation origin of the film pattern.

The surface composition of polymer blends has been intensively investigated during the last decade [342, 349, 358–360]. In most cases, the lower surface energy component concentrates at the free surface to minimise overall free energy. Anomalous behaviour was observed for a blend of high-M_w polystyrene and low-M_w polymethyl methacrylate [300]. PMMA preferentially segregated at the air-polymer interface though it had a higher surface energy. This phenomenon was explained by the effect of the chain end groups and the entropic penalty of the surface adsorbed chains. A combination of SFM techniques was used to study the surface topography and probe the viscoelastic properties [349].

3.2.6
Wetting of Chemically Heterogeneous Substrates

Recently, chemically patterned surfaces have attracted a lot of attention due to their potential use as templates for lateral ordering of polymer films. On a micrometer scale, liquid dewets such surfaces and segregates on surface areas which exhibit preferential interaction with the liquid [360, 361]. A few successful attempts have been made to transfer a lateral variation in surface energy into a composition pattern in the polymer film [16, 362, 363]. Figure 39a shows a laterally patterned structure which consist of periodic stripes of alternating PVP and PS domains. One of the domains, e.g. PVP, could be removed by dissolution in a

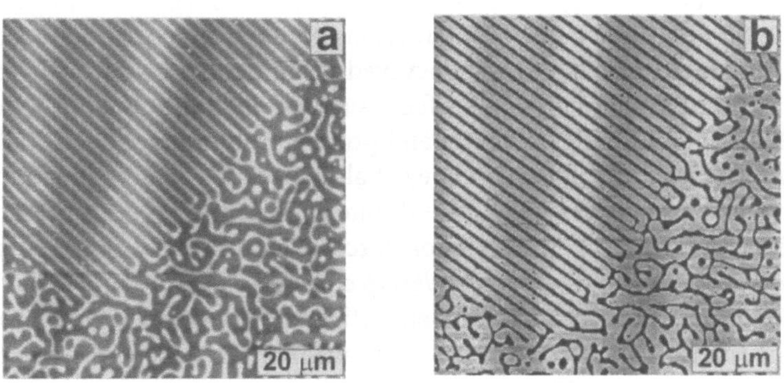

Fig. 39. SFM images of a PS/PVP blend (50/50 wt %) spun cast on a patterned Au substrate [16]. The lateral phase pattern was created by microcontact printing using a polydimethylsiloxane stamp soaked in a octadecyl mercaptane (ODM) solution. The resulting Au/ODM pattern (2.4-mm periodicity) features stripes of alternating surface energy with only little topographic variation. The topographic images show 80×80 μm^2 area of (**a**) as cast and (**b**) after removal of PVP by dissolution in ethanol. The PS and PVP phases are laterally separated where PVP lies on the bare AU surface regions displacing the PS onto the ODM stripes. Courtesy of U. Steiner

selective solvent to yield a PS grating with 2.4 mm periodicity and 65 nm height (Fig. 39b). Here, one made use of two thermodynamic effects: phase separation and selective wetting. The dewetting behaviour becomes less obvious when the chemical pattern is commensurable with the molecular size [364].

3.3 Measuring Interfacial Interactions

Macroscopic properties of heterogeneous systems such as colloidal suspensions, biological cells, thin coatings and blends depend on surface forces that act between the phase boundaries. In addition to adhesive forces in contact, many practical applications, where particles are kept at a certain distance from each other, require detailed information about distance variation of the surface forces. Depending on the surface separation, different forces can be dominant. Van der Waals forces, which determine adhesion, wetting and friction properties of the interface, decay with separation over ca. 10 nm. The decay results in a certain surface pressure which determines stability of thin films and drive spreading of liquids [297–299, 309, 310]. Long range forces such as electrostatic double layer forces, hydrophobic forces and solvation forces might be active at 100-nm distances and control aggregation of colloids in a liquid medium. Special polymer forces – entropic in origin – arise whenever two surfaces bearing a polymer layer are brought into contact [77, 301, 365–367]. Such forces can either promote adhesion between dissimilar solids or prevent aggregation of colloidal particles.

The distance dependence of the normal and shear forces acting between surfaces can be studied using the surface force apparatus and the scanning force microscope. SFA measurements of interactions between adsorbed polymer layers and protein molecules have been surveyed in numerous review articles [368–372]. Different types of polymer surfaces such as block copolymers, tether polymer chains, amphiphilic polymers and polyelectrolytes have been studied regarding interfacial interactions. Kilbey et al. showed how confinement between two surfaces caused ordering of PS-*b*-PI into a lamellar phase in a presence of a non-selective solvent [373]. Depletion forces were observed in nonadsorbing polymer solutions above a certain overlap concentration and molecular weight [374, 375]. Transitions between the attractive and repulsive regimes were observed in different polyelectrolyte systems as function of the ionic strength and the charge density [376, 377].

Complementary to the SFA experiments, SFM techniques enabled direct, non-destructive and non-contact measurement of forces which can be as small as 1 pN. Compared to other probes such as optical tweezers, surface force balance and osmotic stress [378–380], the scanning force microscope has an advantage due to its ability in local force measurements on heterogeneous and rough surfaces with excellent spatial resolution [381]. Thus, a force-distance dependence measured from a small surface area provides a microscopic basis for understanding the macroscopic interfacial properties. Furthermore, lateral mapping of long range forces gives an additional means for visualisation of the surface morphology [93, 99].

Quantitative evaluation of a force-distance curve in the non-contact range represents a serious experimental problem, since most of the SFM systems give deflection of the cantilever versus the displacement of the sample, while the experimentalists wants to obtain the surface stress (force per unit contact area) versus tip-sample separation. A few prerequisites have to be met in order to convert deflection into stress and displacement into tip-sample separation. First, the point of primary tip-sample contact has to be determined to derive the separation from the measured deflection of the cantilever tip and the displacement of the cantilever base [382]. Second, the deflection can be converted into the force under assumption that the cantilever is a harmonic oscillator with a certain spring constant. Several methods have been developed for calibration of the spring constant [383, 384]. Third, the shape of the probe apex as well as its chemical structure has to be characterised. Spherical colloidal particles of known radius (ca. 10 μm) and composition can be used as force probes because they provide more reliable and reproducible data compared to poorly defined SFM tips [385].

Based on these conditions, SFM can be used to measure the work of adhesion, the surface energy, and the non-contact forces between surfaces. Based on the JKR theory, both the work of adhesion and the surface energy could be estimated rather precisely from the pull-off force [386] and the surface deformation [387]. For non-polymer samples, direct correlation between adhesive forces and frictional forces were established for surfaces with different chemical compositions [388–391]. Adhesion of viscoelastic polymers exhibits dependence on the withdrawing rate while the hysteresis between the load and unload curves indicates that more force is required to separate the viscoelastic phases than to bring them into contact [5, 117]. SFM equipped with a colloidal probe has been used to investigate adhesion between glass surface bearing PEO layers [392]. A pronounced long range attraction has been observed on separation of the surfaces up to 100 nm (Fig. 40). Depending on the molecular weight and the degree of coverage the work of adhesion varied up to 60 $\mu J/m^2$.

A variety of long range forces emerges when either two polymer surfaces are separated in a non-polymer liquid, or non-polymer surfaces are separated in a non-adsorbing polymer liquid. In the first case, the interaction depends on the dielectric properties of the liquid and the solubility properties of the polymer. In a poor solvent, van der Waals forces can be measured at short distances as a function of intervening media. The force between a gold sphere attached to the tip and a PTFE surface, could be switched over from attractive to repulsive by changing the polarity of the liquid from polar to less polar, respectively [393, 394]. In good solvents, the steric repulsion as well as bridging forces between brush-like polymer layers have been studied by SFM [135, 395–399]. Both in-phase and out-of-phase force responses of grafted polystyrene chains were measured in solvents of different quality, such as toluene (good) and water (poor) [135]. In toluene, the chains formed a stretched brush whose elastic modulus was an order of magnitude larger than of the collapsed layer in water. Scal-

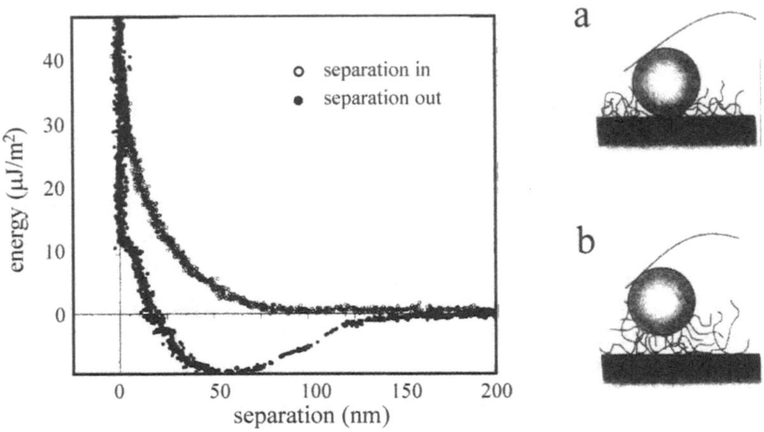

Fig. 40. right: Schematic representation of the polymer layer probed by a spherical particle. Here the particle can be stuck to the surface directly (**a**) or, at greater separations, held by a bridging interaction (**b**). **left:** Interaction energy between a bare glass particle and a glass bearing an adsorbed layer of poly(ethylene oxide) (M_w=205 K) versus surface separation. Note the evidence of bridging adhesion, but the absence of any attraction on approach of the surfaces. Inward (○) and outward (●) runs are plotted. Reproduced from [392]

ing analysis and computer simulation have been used to model the interaction between a layer of end-grafted polymers and a SFM tip upon reducing the solvent quality [397, 398]. In addition, collapse of polymer brushes has been monitored by analysis of the cantilever noise [399].

Electric double layer forces between polyelectrolyte and non-polymer surfaces in aqueous media have also been studied very intensively [371, 394, 400–402]. The adhesion between polyelectrolyte surfaces could be reduced considerably by increasing the ionic strength of the medium [400]. Using an electrochemical cell and a gold coated tip, the adhesion between electroactive layer of poly(vinylferrocene) was controlled through the selective oxidation or reduction of the polymer films [401].

When a polymer solution is placed between non-adsorbing surfaces different forces occur depending on the chain length, the polymer concentration and the dielectric properties of the solvent. Depletion forces arise in dilute solutions where polymer coils become excluded from the interplane compartment and result in osmotic pressure favouring surface attraction [403]. A main technological concern in this case is the ability of the depletion forces to cause aggregation of colloids [378]. In moderately concentrated polymer solutions, oscillatory structural forces can be observed as the tip-surface separation commensurate with the molecular dimensions [404]. The oscillations owing to molecular ordering show up clearly in the MAC mode amplitude curves in Fig. 41 [183]. Similar to simple liquids [180, 181, 405, 406], the geometric confinement caused a local ordering of polymer coils which gave rise to an oscillatory force variations [77, 407]. The least understood and most puzzling phenomena of the re-

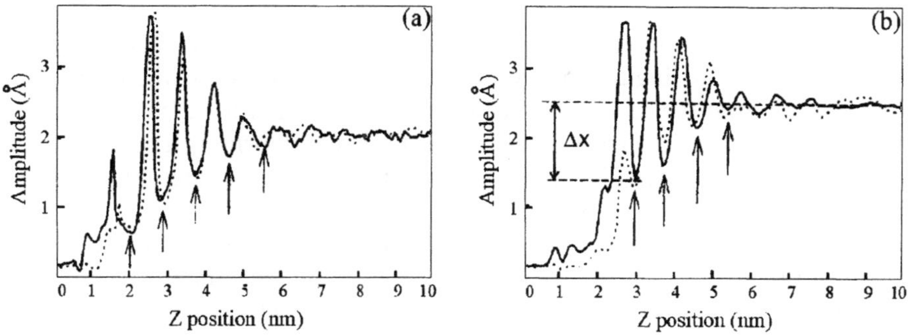

Fig. 41. Two independent measurements of force spectra of the MAC mode SFM at OMCTS-graphite interface. The amplitude of oscillation of the magnetic cantilever driven by an external magnetic field oscillates in both approaching (*solid line*) and retracting (*dotted line*) curves in the region of a few nanometers away from the surface due to ordered layers of OMCTS molecules at the interface. The period of oscillation 8.2 Å reflects the dimension of OMCTS molecules along the direction perpendicular to the layers. **a** Driving frequency 500 Hz, scan rate 2.8 nm/s. **b** 200 Hz and 1.6 nm/s. The *arrows* on the plots correspond to repulsive-force maxima. Reproduced from [183]

cent years concern the attractive interaction between hydrophobic surfaces in a polar medium. Ducker et al. has used the scanning force microscope to determine an excess of attractive forces over electrostatic double layer forces between an air bubble and a silica particle [408]. The additional attraction was shown to be active at distances up to 50 nm and ascribed to the long range hydrophobic forces.

3.4
Micromechanical Properties

Micromechanical properties of polymers by SFM can be studied using two different approaches. In the first one, the scanning force microscope is used to examine structure of either ex-situ [84, 88, 409, 410] or in-situ deformed samples [411]. In these experiments, the tip was used as a *"passive probe"* to obtain submicrometer information about structure changes caused by uniaxial stretching of amorphous polystyrene [409], ultra-high-molecular-weight polyethylene [84, 410], and hard elastic polypropylene [411]. In the second and more interesting approach, the SFM tip has been used as an *"active probe"* which both deform the sample and simultaneously measures the reaction force as well as the induced deformation [67, 70, 115, 122, 136, 159, 412, 413]. Thus SFM can investigate mechanical properties at interfaces and provide microscopic basis for the understanding of friction, lubrication, viscoelasticity, indentation and wear.

In the active probe approach, SFM acquires both static and dynamic mechanical properties (Sect. 2.2.2). The former includes the shear and Young's modulus (G,E) as well as the surface indentation and contact area (δ,a). Dynamic measurements are able to discriminate the elastic and damping properties (E′, E″, tgφ) and study the time dependence of the modulus and compliance. In addition, one has to distinguish between mechanical experiments when the tip is pulled away from the surface or pushed towards to the surface (Fig. 12). In the first case, the tip can stretch a small portion of material forming a neck between the tip and the sample, while in the second case, the tip penetrates into the sample and probes compressibility.

When the experimentalist set an ambitious objective to evaluate micromechanical properties quantitatively, he will predictably encounter a few fundamental problems. At first, the continuum description which is usually used in contact mechanics might be not applicable for contact areas as small as 1–10 nm [116, 117]. Secondly, since most of the polymers demonstrate a combination of elastic and viscous behaviour, an appropriate model is required to derive the contact area and the stress field upon indentation a viscoelastic and adhesive sample [116, 120]. In this case, the duration of the contact and the scanning rate are not unimportant parameters. Moreover, bending of the cantilever results in a complicated motion of the tip including compression, shear and friction effects [131, 132]. Third, plastic or inelastic deformation has to be taken into account in data interpretation. Concerning experimental conditions, the most important is to perform a set of calibrations procedures which includes the (x,y,z) calibration of the piezoelectric transducers, the determination of the spring constants of the cantilever, and the evaluation of the tip shape. The experimentalist has to eliminate surface contamination's and be certain about the chemical composition of the tip and the sample.

Micromechanical experiments made so far can be roughly divided into two parts: (i) design of special techniques to measure and evaluate separately different contributions in the net force, such as adhesion, friction, deformation, and (ii) imaging of various heterogeneous surfaces such as blends, composites and microphase separated structures by conventional SFM's to collect statistical information and understand the origin of the mechanical contrast. Many of the micromechanical experiments were discussed elsewhere [58, 67, 68, 381, 412–414]. Here we will focus on recent advances in analytical applications of the "active probe" SFM.

The force-distance curve in Fig. 12 contains a complete set of information about the static mechanical properties of the sample. When the deformation ε and the force $F=k_c\delta$ are determined one can use a continuum model from contact mechanics to estimate Young's modulus. Assuming small contribution of the surface forces, the Hertzian model has been chosen to evaluate modulus as a function of the indentation depth for different polymers such as PS, PVC, PU and PI rubber [415]. Although the model overestimated absolute values of modulus, it showed good agreement with elastic bulk moduli which differed by 4 orders of magnitude for the set of polymers. The best correlation was found at large indentation depth in the range 20–200 nm (Fig. 42). From the hysteresis of the force-distance curve, the dissipated energy and the inelastic deformation can be measured and correlated with the activation energy of the segmental mo-

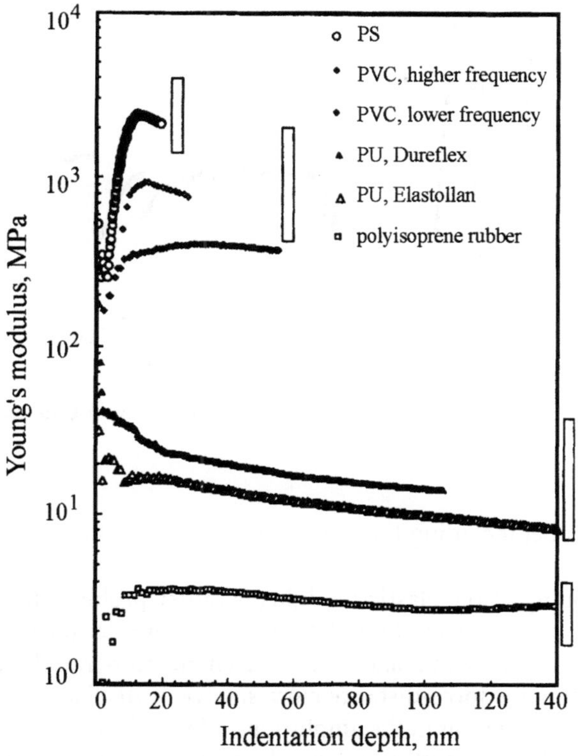

Fig. 42. Depth variation of the elastic modulus for rubber, two different polyurethanes, PVC and PS as a function of the indentation depth. Bars demonstrate the range of elastic bulk modulus variation for a specific material. Reproduced from [415]

tion in the polymer chain [416]. At small deformations, however, the modulus exhibited strong variations with the indentation depth within one order of magnitude (Fig. 42). This result is unusual and has to be carefully interpreted as the continuum theories might be not valid in the limit of small contact areas. Other intractable effects are thermal drift and stress relaxation which can take part because of the relatively long acquisition time (~1 Hz). Besides the force-distance experiments, contact mode SFM was used to study creep of PTFE [417], compliance of diblock copolymer films of PVP-PS depending on the solvent selectivity [135] and viscoelasticity of human platelets [418].

In addition to the compression loading, uniaxial extension of entangled PDMS chains have been investigated by pulling a small portion of the material and measuring elastic response before the rupture happens [419]. The multiple ruptures observed in the force-distance curves (Fig. 43) have been interpreted as fractures of an entangled network of PDMS chains formed between the tip and the silica grafted surface. At small deformations, also the capillary forces were shown to contribute in the force. The elastic part of the curves was described us-

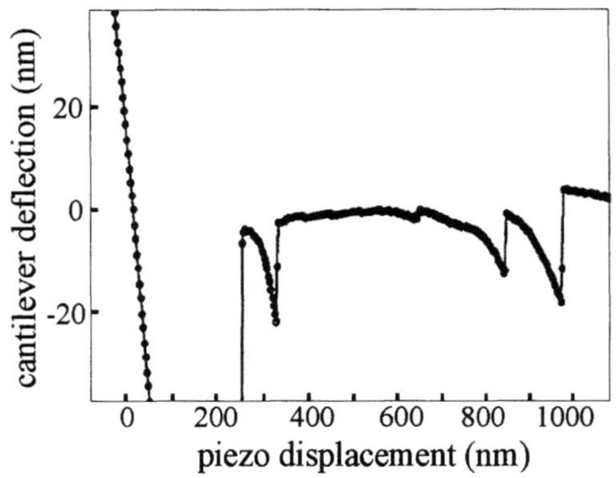

Fig. 43. Example of multiple rupture obtained with polydimethysiloxane grafted onto silica [419]

ing the tube model including the inextensibility of the chain at high deformation.

Lateral mapping of the elastic modulus is rather problematic as it requires significant data storage space. Modulation techniques offer the possibility to reduce data storage space and make the measurement more rapid (Sect. 2.2.2). Instead of recording the force-distance curves, force or displacement modulation SFM acquires the variation in the amplitude and the phase lag. Below the cantilever resonance, the amplitude is independent on the cantilever mass and directly related to the contact stiffness k_s. However, because the contact stiffness is a product of the elastic modulus and the contact area which is usually is unknown parameter, FMM-SFM can only discriminate qualitatively between more stiff or more compliant regions on the surface [118, 124, 125, 420, 421]. This capability was demonstrated on a variety of filled and unfilled industrial blends [128]. In combination with cryogenic facing, force modulation SFM was used to investigate microstructure of blends containing isobutylene, polypropylene, polyisoprene and carbon black (Fig. 44). Special attention deserves dynamic viscoelastic measurement of liquid PDMS films constrained between a flat silicon wafer and a glass sphere [127]. The frequency spectrum and even the microscopic viscosity of PDMS were evaluated using a mechanical model assuming the Newtonian behaviour of the liquid.

As discussed in Sect. 2.2.2, FMM images can lose the material contrast when the sample stiffness exceeds the stiffness of the cantilever. In addition, the net signal contains friction effects because of the cantilever bending and the sample indentation. Furthermore, in liquid samples, capillary forces dominate the response at low frequencies [127]. These drawbacks can be overcome by operating the microscope above the contact resonance frequencies. In the so-called contact-mode scanning local-acceleration microscope the cantilever oscillates at very low amplitudes of ca. 0.1 nm which still provides strong enough contrast with respect to the mechanical properties [122]. Since the response of the canti-

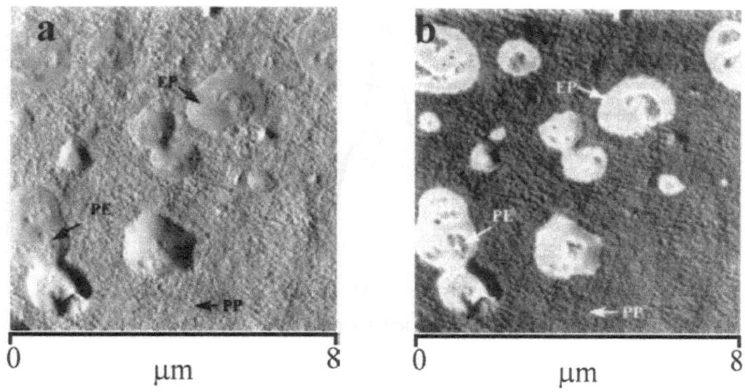

Fig. 44. The SFM amplitude (**a**) and force modulation (**b**) maps of a cryogenic faced impact copolymer (ICP) composed of a polypropylene (PP) matrix with high ethylene (60 wt.%) ethylene-propylene copolymer (EP). Crystalline polyethylene (PE) phases are seen in the EP domains, which are surrounded by the PP matrix. Modulus contrast in the force modulation (drive amplitude 100 mV) image associated with the three polymers: the stiff PP matrix is *dark*, the soft EP domains are *light*. The crystalline PE regions have modulus between the PP and the EP, thus an *intermediate shade of grey* is observed for the PE domains [128]

lever does not depend on the cantilever stiffness, it provides direct information on the contact stiffness and the interaction damping.

Recently, the SLAM technique was modified to make local mechanical spectroscopy as a function of temperature [138]. Figure 45 shows phase-lag and amplitude spectra of polyvinylchloride (PVC) around its glass transition temperature. The drop in the amplitude and the extremum in the phase lag indicate changes in the elastic and damping properties of PVC caused by the glass transition. In addition to local spectroscopy, the contact-mode SLAM enabled lateral mapping of both topography and sample stiffness. Figure 46 shows SLAM images of a PVC/PB polymer blend at room temperature and above the glass transition of PVC. At the lower temperature, the stiffness of the PB inclusions and of the PVC matrix is approximately the same, while the inclusions protrude above the surface. At the elevated temperature, the matrix became more compliant and the inclusions became embedded into the matrix. The change in stiffness was explained by decrease of the elastic modulus of PVC above the glass transition temperature. The variations in topography were attributed to the difference in thermal expansion coefficients or preparation history of the sample.

Tapping mode SFM shows the most versatile and strong material contrast due to variations in stiffness, viscosity and adhesion. While the theoretical basis of the tapping mode is still under development (Sect. 2.2.2), the results obtained so far can be used for qualitative discrimination between different phases in a heterogeneous sample. One should keep in mind, that the strong dependence of the contrast on the experimental parameters (ω, A, r_{sp}) may lead to quantitative mis-

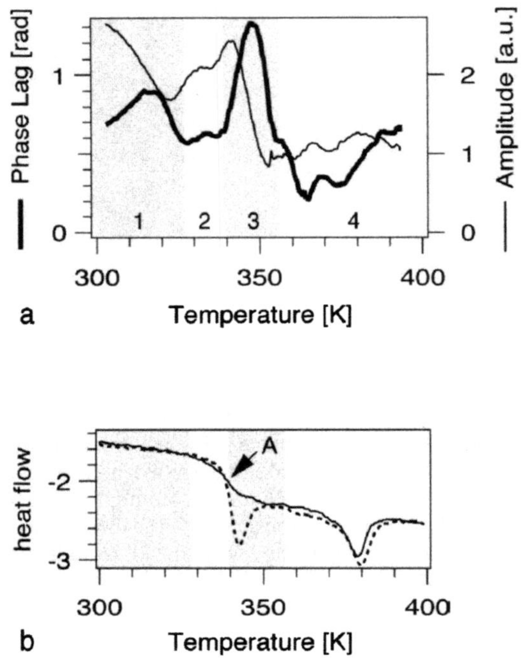

Fig. 45. a Local mechanical spectroscopy spectra as a function of temperature of the technical polyvinylchloride (PVC): the vibration amplitude is displayed as a thin line and the phase lag as a thick line. Four domains (1–4) differing in the signal changes are distinguished on the spectra. **b** DSC measurements of the same sample, where the graph displays the first (*dashed line*) and the second (*solid line*) heating. The glass transition can be recognised around 340 K (labeled *A*). A reversible endothermic peak around 380 K might correspond to the crystalline phase. Reproduced from [138]

interpretation of the amplitude and phase signals and even cause inversion of the contrast [109–112]. In particular for viscoelastic materials, viscosity and elasticity are two independent properties which act simultaneously upon tapping and can lead to various crossover phenomena [102, 107]. Figure 47 shows the height and phase images recorded for a microlayered PE sample at different r_{sp}. On going from light ($r_{sp}\approx 0.8$) via moderate ($r_{sp}\approx 0.5$) to hard ($r_{sp}\approx 0.2$) tapping, the topography images do not show much difference but the phases do. The phase contrast can reverse at hard tapping indicating that the overall force might change its character from repulsive to attractive (Fig. 47c).

Complementary to the indentation experiments, the lateral force microscope has been introduced to measure lateral forces exerted on the tip [388]. Somewhat latter, different modifications of contact-mode SFM have been developed to investigate indentation and wear in thin films [115]. On the lateral force measurement, one has to distinguish between operation at very small lateral displacements of the cantilever when the shear forces dominate in the net force, and a

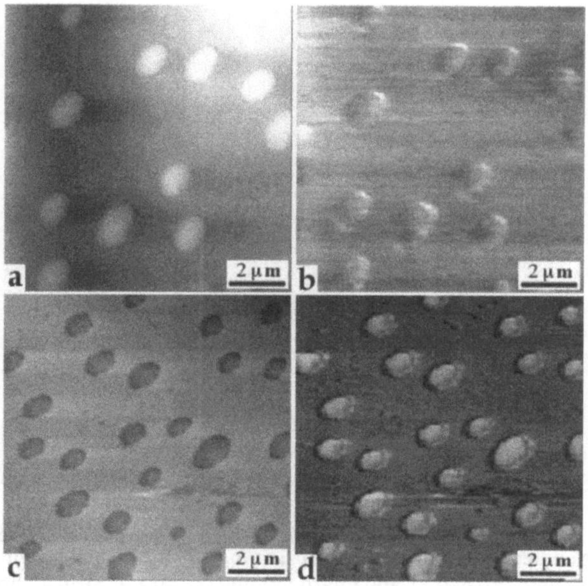

Fig. 46. Normal force (topography) images (**a**,**b**) and SLAM amplitude (stiffness) images (**c**,**d**) of a PVC/PB polymer blend at room temperature (**a**,**b**) and at 373 K (**c**,**d**). At high temperature, the matrix became more compliant (*darker* in the SLAM image). Reproduced from [138]

conventional regime when the tip slides over the studies and probe the friction properties. The friction force can be different in origin and usually results from a combination of different effects such as adhesion, viscoelastic deformation and surface roughness.

This discipline has become known as nanotribology and studies friction, adhesion, lubrication and wear on a scale of few nanometers down to the atomic scale [115–119, 412, 422, 423]. Most of the studies have been done on bare substrates, such as Si wafer, graphite, mica, or model surfaces, such as SAM's and LB films, where the microscale friction is generally found to be lower than the friction on macroscales due to smaller contribution of ploughing in the microscale measurements [424]. Depending on the load, the microscale friction approaches macroscopic values at high contact stresses, while wear is negligible at low forces if they are below the yielding and fracture points. In hard samples such as ceramics and Si wafers, the material was shown to be removed by the ploughing mode in a brittle manner without much plastic deformation [425].

In comparison to metals and ceramics, much less attention has been directed to friction properties of polymers. The main reason for this is that the use of polymers in machinery is not so common. However, recent progress in polymer technology has enabled partial substitution of metals by polymeric materials in addition to the wide utilisation of polymer based lubricants. When a solid material is allowed to slide over a polymer surface, complicated factors other than

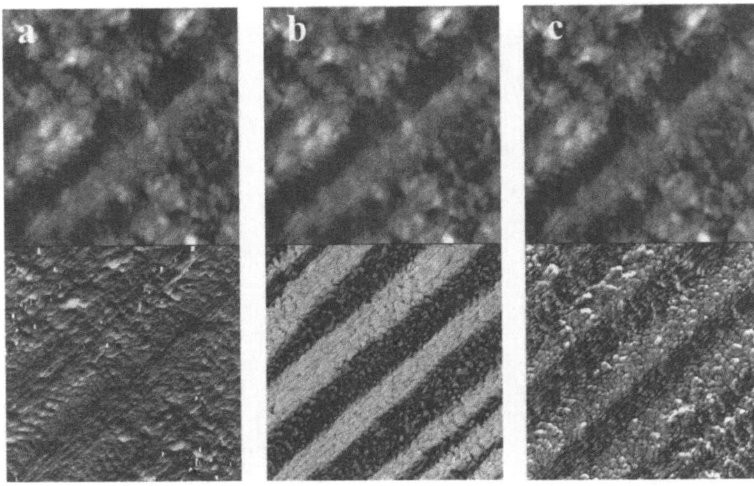

Fig. 47. Height (*top*) and phase (*bottom*) images of a microlayered PE sample recorded on the surface perpendicular to the layer plane using A_0=100 nm: **a** light tapping (r_{sp}=0.8–0.9); **b** moderate tapping (r_{sp}=0.4–0.7); and **c** hard tapping (r_{sp}=0.1–0.3). In all height images the contrast covers variations in the 0–170 nm range. In the phase images, the contrast covers phase shifts in the 20° (**a,b**) and 50° in (**c**). Reproduced from [110]

those for rigid materials may arise [4]. The instability of the polymer surface, strong dependence of the surface properties on the environmental conditions, as well as the peculiar dehesion/readhesion waves make the investigation of friction properties more difficult.

LFM of thin gelatine films revealed an entire spectrum of tribological behaviour depending on the scanning velocity and the applied load [426]. Friction and wear behaviours were distinguished between the glassy, the rubbery and the melt phases. Of fundamental importance was the finding that energy, transferred to the film, caused local heating which melted the crystalline phase and induced rubbery behaviour on the amorphous phase. Recently, cross sections of microphase separated block copolymers of polystyrene and poly(2-vinyl pyridine) have been investigated by LFM regarding friction coefficients depending on the chemical structure of the tips [427]. A few studies have been performed on thin polymer layers with regards to their lubrication performance [115, 428, 429]. The friction coefficient of a Si wafer coated with an ultrathin film of PFPE was usually lower than an unlubricated surface, however, depending on the coating structure the lubrication properties were shown to degrade during long term operation.

3.5 Surface Patterning and Nanofabrication

Today the frontier of the fabrication of electronic devices has moved from the micrometer scale down to tens of nanometers scale. Scaled-down conventional devices such as field-effect transistors and devices based on quantum effects are two most prominent examples of the electronic miniaturisation [20, 23, 430]. The major challenges in preparation of such devices are (i) growing the substrate materials and (ii) patterning the substrates. Whereas the former rely on self-organisation of the surface structure, the substrate patterning on the nanoscale requires special tools.

Shortly after the invention of STM and SFM, it was noticed that the scanned surface area revealed different morphology depending on the scanning conditions. This observation opened a new field of applications of SPM techniques directed towards surface modification and data storage on the nanoscale and nowadays called scanning probe-based lithography (SPL) [414,431,432]. Most interesting examples of SPL capabilities include single molecule manipulation [433–436], electrochemical deposition of metal nanoclusters [437], local polymerisation [438], and field emission modification of polymer resists [439,440].

In comparison to STM, robust SFM utilities are less suitable for nanolithography unless they are specially modified. First, electronics must be open for generation of the pulsed bias (or force) and synchronisation of this pulse with the tip motion. The sample stage should be equipped with a micrometer positioning system in order to enable analysis of previously produced structures. Special tips may be required with a high durability, hardness and sufficient electrical conductivity. Depending on the mechanism of surface modification one can distinguish electrical, mechanical, chemical, and optical SPL methods. However, not all these methods can be implemented for nanostructuring of polymer surfaces.

3.5.1 *Electrical Lithography*

An electric field applied between the tip and sample gives us a powerful means for surface modification including electrochemistry, electrostatic charging, or merely ohmic heating. Usually, electrical modifications do not lead to profound changes in the surface topography. Moreover, bias pulses enable modifications on smaller scales compared to a mechanically induced pattern. Most of the SFM based electrical lithography have been done on inorganic substrates such as hydrogenated silicon wafers [441–445]. The terminal hydrogen atoms can be locally removed by applying a bias pulse so that the stripped areas became susceptible to oxidation. When the oxidation is performed in the ambient, the pattern quality strongly depends on humidity. Figure 48 shows oxide stripes on a Si (100) sample, where the stripe width could be reduced from 90 nm to 4 nm by lowering the relative humidity from 61% to 14% [445].

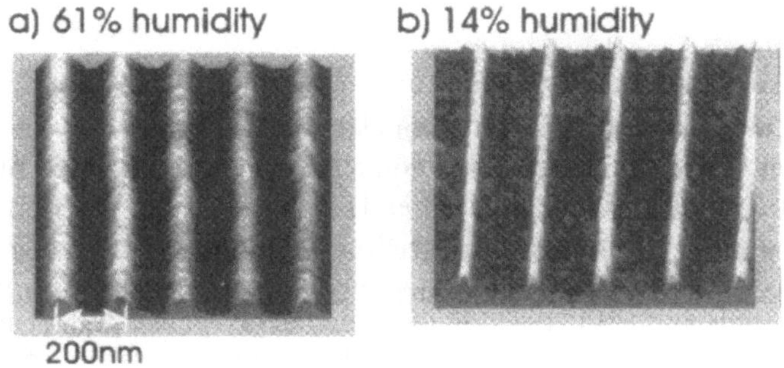

Fig. 48. The surface patter of SiO_2 stripes on a silicon wafer was prepared electrochemically by applying a bias pulse to locally remove the terminal hydrogen atoms. The aspect ratio (height/width) of oxide lines improves significantly when the relative humidity is lowered from 61% to 14%. Reproduced from [445]

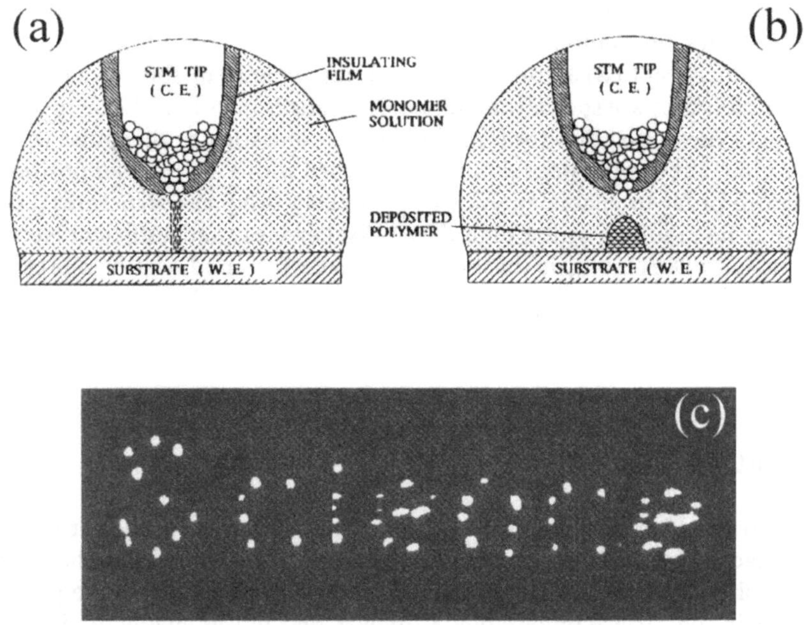

Fig. 49. A schematic to illustrate the basic concept of the deposition process. **a** The insulated tip serves as the counter electrode and the substrate serves as the working electrode in the electrochemical cell. **b** A small amount of polymer will be directly polymerised onto the substrate following a single high voltage pulse. **c** An STM image of the word "science", which was written by polymerising polypyrole directly onto a gold surface. The size of the polymer dots are 5–10 nm across. Reproduced from [453]

Because of some apparent restrictions, organic materials including polymers are not very suitable for electrical SPL. First, the surface structure of organic materials is less stable and less defined compared to inorganic solids. Second, the materials to be used for electrical SPL have to be either electrically or optically active. Most of the SPL experiments have used Langmuir-Blodget films and self-assembled monolayers (SAM's) due to their well defined thickness, orientation of the molecules and uniform surface composition [446–450].

In addition to the conventional lithographic techniques, surface patterning was performed by means of local polymerisation of the monomers under the SPM tip. These studies have been mainly focused towards electrically conductive polymers such as polypyrole, polythyophene and polyaniline. The easiest way to implement polymerisation is to set either the tip or sample potential sufficiently positive to cause the electrochemical oxidation of the monomer [438, 451–455]. This technique enabled controlled removal and deposition of polymer dots as small as 1 nm to in a well defined pattern [453]. After deposition, the dots could be read using a conventional imaging mode (Fig. 49).

Also polymers which are not lithographically active have shown some useful properties. STM has been used to produce from 0.5 to 2 nm sized structures in thin films of polyvinyl alcohol [456]. The structures were shown to depend strongly on the relative humidity and their origin might involve electrochemical reactions.

3.5.2
Mechanical Modification

As compared to electrical SPL, a broader range of materials becomes accessible for surface modifications using mechanical means. Two different approaches have been developed based on SFM: surface scratching and material transfer. In the first approach, a scratch or trench of a defined width and depth can be produced by dragging the tip across the sample under controlled load and speed. In the second and more elegant approach, small portions of the sample, e.g. molecules and particles, are picked up by the tip, lifted and transferred to another region of the surface.

Different types of lithographic constructions, such as lines, grooves and pits, could be inscribed by a SFM in thin metal films [457–459]. In a similar way, SFM has been employed to scratch thin polymer coatings for the photoresist application. Depending on the mechanical properties of the polymer, the applied forces were varied from 5 to 300 nN to produce grooves with a width around 30 nm and a depth of ca. 5–10 nm [460–462]. Two patterning modes have been identified depending on the applied load. Below a certain force value, the produced pattern appeared raised, whereas at higher forces the material was removed and piled up at the edges of the grooves [460, 463]. Different deformation mechanisms were also observed while studying polymer blends of polystyrene and polyethylene oxide, where the PEO component revealed greater compliance [420]. The deformation mechanism was investigated in terms of the energy dissipation and inelastic deformation induced by the SFM tip when scanning a polymer film [416].

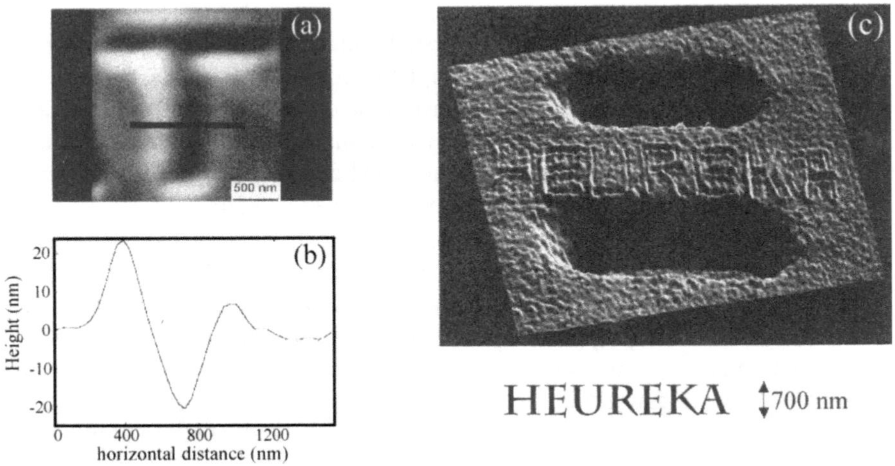

Fig. 50. a SFM picture of an indentation of the letter "T" on PMMA with a nickel tip. Force used for the imaging was 10^{-6} N. **b** Profile along the line drawn in (**a**). Reproduced from [416]. **c** "HEUREKA" written with the dynamic ploughing technique on a compact disc (CD) of polycarbonate. The holes are information pits in the disc. The letter height is 700 nm, and the indentation depth is 10 nm. Reproduced from [464]

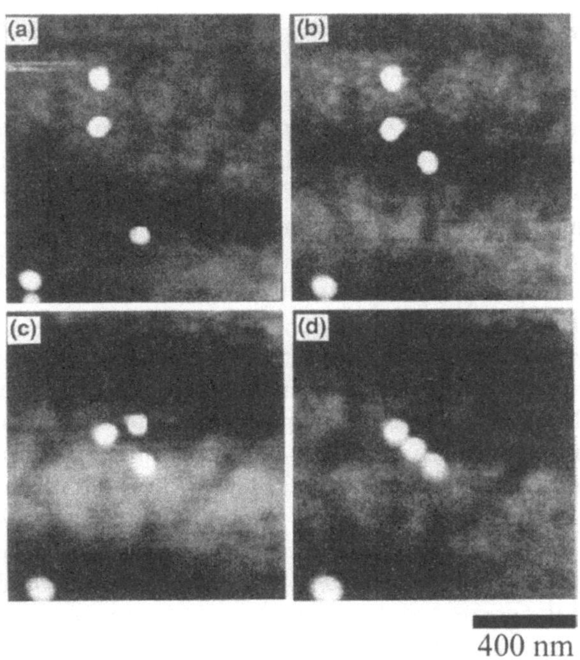

Fig. 51. A sequence of SFM images acquired during the formation of GaAs nanoparticles on a GaAs surface. The SFM tip was used to push the particles to their final positions in the line. Reproduced from [468]

Under well controlled conditions, a micrometer letter "T" in Fig. 50a was produced in a thick PMMA film by 75 nm indentation of the tip into the sample which resulted in a trench depth of 20 nm.

Recently, dynamic mode SFM has been used for gen1tle and more controlled surface modification [464,465]. Depending on the substrate type, the loading force was varied with ca. 500 nN amplitude at a frequency of 0.1–10 kHz. Force modulation SFM was employed to write on polycarbonate surfaces with letters ca. 50 nm high (Fig. 50b) as the tip was in light contact with the surface and a modulation frequency was relatively low to ensure plastic deformation of the polymer [464].

A few other examples have to be mentioned though they are not polymer related. As an alternative to scrapping out a portion of a surface, Butt et al. has proposed to use a SFM tip as a carrier of molecules which can be selectively adsorbed to a defined surface region [466]. With this approach, SAM patterned surfaces can be performed on much smaller scales compared to the μCP technique [23, 213]. In the other mechanical approach, SFM tips have been used to push (move) single atoms or small clusters on a surface in a defined arrangement [467,468]. Figure 51 shows a sequence of SFM images captured during formation of a line of 30 nm GaAs nanoparticles [468].

3.5.3
Chemical Synthesis on the Nanoscale

As compared to electrical and mechanical SPL's, SPM based approaches for the chemical synthesis of nanostructures are still in their infancy. Recently, highly localised chemical catalysis was carried out on the surface of a self-assembled monolayer with a platinum-coated AFM tip [469]. The terminal azide groups of the monolayer were catalytically hydrogenated and converted to amino groups (Fig. 52). In a second step, the amino groups formed by this process could be selectively modified with a variety of reagents to generate more complex structures with high spatial resolution. The use of a catalyst rather than an electrochemical process allows the use of a variety of SPM devices and surfaces, and requires no potential between the tip and sample.

4
Visualisation and Probing of Single Macromolecules

Modern nanotechnology and nanofabrication processes are advancing towards manipulation of structure and functions on the molecular scale [19–23]. Many innovations and strategic areas of research which have appeared during the last 5–10 years corroborate the famous opinion of R. Feynman "there is a plenty of space on the bottom" [471]. Some prominent examples include self-assembly of small molecules and their ordering at interfaces [8, 220, 472, 473], three-dimensional macromolecules with a defined shape, interior and surface structure [34–38], and templating of biomolecules [39–41, 474]. Most of these concepts follow a biomimetic approach, where synthetic structures mimic organisation princi-

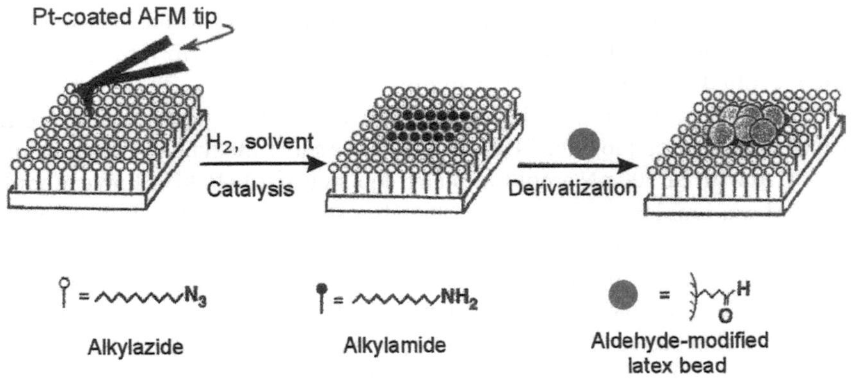

Fig. 52. Scanning with a platinum coated SFM tip over a SAM surface containing terminal azide groups in the presence of H_2 leads to the reduction of azide groups to primary amino groups. Derivatization of the resulting amine surface with aldehyde-modified latex beads results in specific labelling of the reduced areas. Reproduced from [469]

ples of biological objects and rely on the kinetic ability of the system to reach its thermodynamic equilibrium. An alternative approach is constructing "things" with single molecules, which combines a bottom-up approach based on molecular self-organisation and a top-down approach using special techniques which enable manual control over a desired pattern. There is an endless list of synthetic concepts and key words which have recently emerged based on extraordinary progress in understanding of the structure and dynamics of single molecules: molecular motors, molecular muscles, molecular templates, molecular reactors, and molecular electronics. The progress in these fields is greatly assisted by scanning probe techniques which can identify objects as small as 1 nm, position them on a surface and probe their physical properties.

4.1 Biological Molecules and Colloids

Because of the nanoscopic dimensions (1–100 nm) and structural complexity of the biological objects [39], the number of tools available for the biochemist is significantly more limited compared to synthetic and inorganic materials. Some of the structures are too complex for X-ray and NMR, and too small for optical microscopy. For many years, biologists have been looking for a complementary technique which would combine high resolution with operation in a physiologically relevant environment. This became possible with the invention of the scanning force microscope whose resolution of biological samples has approached 10 Å, which is as good as of the transmission electron microscope. Currently, 3D reconstruction of the molecular structure (X-ray) and identification of specific atoms (TEM) in combination with the live observation and probing of single molecules (SFM) represents a powerful approach in biophysics.

The frontier of biological SFM has advanced through a series of stages [48, 62, 475–478]. At the beginning, main efforts have been made to develop methods for reliable deposition and visualisation of single macromolecules. Subsequently, the SFM has been used to characterise the structure of various biomolecules and biocolloids, such as nucleic acids, proteins, chromosomes, viruses and cells, as well as nucleo-protein assemblies. Recent studies focus on in-vitro investigation of real time dynamics, specific interactions and manipulation of biomolecules. In order to keep this review relevant to synthetic polymers the following topics are selected for discussion: (i) deposition of single molecules such as DNA onto a solid surface, (ii) imaging conditions for high resolution of their native structure, (iii) statistical analysis of the polymer chain, (iv) interfacial ordering, and (v) dynamics of individual molecules.

Irrespective of the experiment to be done, sample preparation contains a number of necessary conditions. First, aggregation must be prevented if one wants to investigate structure and conformation of single molecules. Second, the adsorption process must be reversible, or at least, very slow in order to approach the equilibrium state and allow statistical analysis of the molecular assembly. Third, adhesion of the molecules to the substrate must be strong enough to sustain the mechanical and adhesive interactions with the tip. However, it should be relatively low to prevent the native structure from deformation.

Compared to other biomolecules, nucleic acids have been studied more intensively, mostly for methodological reasons. At first, nucleic acids have well defined structure which is easily identified amongst other features at the surface. Secondly, their secondary structure is stable and sustain adhesion to the substrate. Different techniques and deposition protocols have been reported to attach biomolecules to a solid substrate and thus enable imaging of DNA under water as well as in air. Most of the techniques make use of electrostatic binding of the molecules to mica. In water, mica is negatively charged [479, 480] and therefore, repels negatively charged DNA molecules. Most common deposition buffers contain up to 100 mM monovalent ions and 1–10 mM of divalent cations, such as Mg^{++}, Mn^{++}, Co^{++}, and Ca^{++} [481]. The divalent cations promote DNA adhesion to the mica surface by inverting its charge [482]. Other methods require treating mica with aminopropyltrimethoxy silane [483] or non-ionic detergents [484]. Low DNA concentrations of 0.5–2 nM are usually used in order to obtain single molecules evenly dispersed over the surface (Fig. 53a). Finally, dehydration of DNA in propanol or in vacuum prior to the measurement favours stronger adhesion of the molecules to mica in the aqueous environment [485].

Strong electrostatic binding to the surface allowed DNA molecules to be scanned for long periods of time without damage and displacement. First reliable images of DNA were obtained in air under low humidity conditions [486, 487] or in propanol [488]. At low humidity, DNA molecules can withstand forces as high as 130 nN. Whereas at high humidity (RH>35%), the molecules can be dissected or swept away by the tip even with a net force of 30 nN [489]. The main cause for that is the capillary forces which vanish upon scanning under liquid.

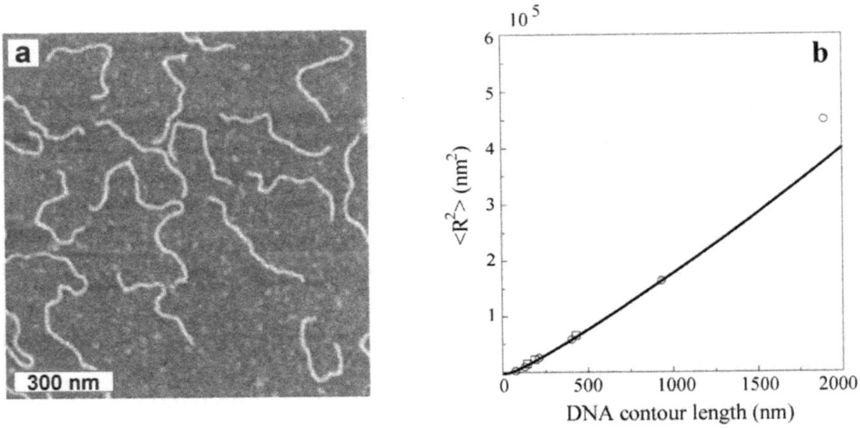

Fig. 53. a DNA molecules of 5994 bp deposited on mica from a physiological solution. **b** Mean-square end-to-end distance, $<R^2>$, as a function of the DNA contour length measured in different buffer solutions [493]. The continuous line represents the $<R^2>$ evaluated from Eq. (10) for a DNA persistent length of 53 nm

For example, imaging forces in propanol were typically one order of magnitude lower than in air resulting in smaller deformations [170]. Moreover, liquids like propanol prevent various contaminants from adhering to the tip during scanning. Another requirement for stable imaging is to use a sharper tip to reduce the adhesion and the capillary forces. Also hydrophobised tips might help to lower the tip-sample interactions.

Imaging possibilities have been enhanced significantly with the introduction of the intermittent contact modes [490–491]. Tapping alleviated the requirement for a strong attachment of the DNA molecules to the surface, since lateral forces were removed and perpendicular forces were reduced considerably due to higher sensitivity of the amplitude to the force variations. The resulting DNA images demonstrated higher contrast than those obtained using contact mode SFM. One of the highest resolution has been achieved for DNA molecules with an average width of the double helix of 3.5 nm using MAC mode SFM [491]. At present, intermittent contact SFM in combination with a sharp tip, controlled humidity and strong attachment to the substrate enables stable and reliable imaging of nucleic acids with a resolution of ca. 3 nm.

When molecules adsorb to a flat substrate, their conformation is modified due to the geometric confinement between the two interfaces and the direct interaction to the substrate. This state can be far from equilibrium if the adsorption process has been fast and irreversible. In this case, the molecules do not have time to sample the whole assembly of thermodynamic states and get trapped kinetically at contact sites. The reversibility is difficult to achieve because of the great size of the molecules and strong adhesion which might exceeds k_BT by far. In order to approach an equilibrium state, the sample has to be prepared under slow adsorption conditions on a time scale comparable to equilibration times of a polymer chain. The equilibration time increases with the chain size, the density of binding sites and the binding energy [481]. Optimal deposition conditions were found, where the transfer of DNA molecules from solution to the substrate is controlled solely by diffusion [493]. For example, for 0.5 nM DNA molecules in a pH 7.4 buffer containing 10 mM NaCl and 2 mM $MgCl_2$, the diffusion coefficient $D=5.5\times10^{-8}$ cm^2/s was found using the formula $n_s / n_0 = \sqrt{4D/\pi}\sqrt{t}$, where n_s - surface concentration of the adsorbed molecules and n_0-concentration of the molecules in solution. Under these conditions, the molecules adsorbed onto mica as single species as shown in Fig. 53a.

Assuming that the deposition process is dominated by equilibration rather than kinetic trapping effects, Rivetti et al. have performed statistical analysis of the chain conformation in terms of the mean-square end-to-end distance, $\langle R^2 \rangle$, and the mean-square bend angle, $\langle \Theta^2 \rangle$ [493]. Both values are well described theoretically for both two-dimensional and three-dimensional states [494–498]. For a two dimensional system, the following formula can be used:

$$\langle R^2 \rangle = 4PL\left[1-\frac{P}{L}\left(1-e^{-L/2P}\right)\right] \rightarrow 4PL \qquad \text{for } L\rightarrow\infty \tag{10}$$

$$\langle \Theta^2 \rangle = \frac{l}{P}$$

where P and L are the persistence length and the contour length of the molecule, respectively, and l is the arc length spanned by the bend angle Θ. Figure 53b shows a plots of $\langle R^2 \rangle$ as a function of the contour length. The slope of the plot could be approximated using a persistence length of 53 nm [499]. The data were in good agreement with those obtained by electron microscopy and gel electrophoresis [500].

Besides nucleic acids, also cell membranes and membrane-bound proteins have been intensively investigated [501–508]. In this case, the SFM measurement is facilitated by the atomically smooth surface of the membranes and strong adsorption to hydrophilic surfaces. High resolution images of membrane bounded *Escherichia coli* were obtained by tapping mode SFM which provided unique evidences to the previously proposed model for ECF_0 permitting internal rotary motions [503, 504]. Complementary, an exceptionally high resolution has been achieved for proteins stabilised by chemical crosslinking using contact mode SFM [505, 506]. Filamentous structures of F-actin with ca. 36 nm periodicity were observed by non-contact mode SFM [507]. Somewhat latter, intermittent contact SFM has been used to image both globular and filamentous proteins in physiological solutions without any visible damage [184]. A helical pitch of 37 nm could be resolved in agreement with the X-ray data. Recently, in-vitro studies of microtubule structure has been able to reveal the individual tubule within one protofilament [508].

In addition to the structure of individual biomolecules, also their aggregation behaviour and interfacial ordering have been investigated. Globular proteins such as ferritin, lysozyme and chymotrypsinogen showed ordered arrays upon adsorption on various substrates [478,509–512]. Recently, a very high resolution of ca. 1 nm has been achieved on imaging of purple membranes in aqueous solutions [511]. The resolution in this case is favoured by the smooth surface structure of the membranes and the mechanical stability of the crystalline array. Also DNA molecules have been shown to condense into various ordered phases at high concentrations in aqueous solutions [513]. Lipid membranes were used as a platform to achieve a close 2D-packing of the molecules (Fig. 54a) which appeared to be independent on the length of DNA and disregarded whether the DNA molecules were circular or linear [514]. In addition to the semiflexible DNA molecules, rod-like particles of cytoplasmic microtubules [515] and the tobacco mosaic virus (TMV) were readily imaged by SFM [516–518]. According to X-ray experiments, TMV rods are 300 nm in length and 18 nm in diameter. The TMV surface has a net negative charge in aqueous solutions under neutral pH. Depending on the ionic strength of the solution and the presence of small detergent molecules, TMV rods either randomly dispersed over the substrate or condensed into a 2D ordered array (Fig. 54b). Different aggregation mechanisms have been proposed, though measurements in liquid are required to avoid capillary forces leading in aggregation [519].

To summarise, recent developments in SFM concerning sample preparation and imaging techniques caused significant progress in biological sciences by making it possible in-vitro visualisation of single molecules in their natural environment with a resolution as high as 10 Å. This inspired very exiting experiments such as (i) real time diffusion of DNA [520–522] and motion of proteins [512, 523]; (ii) in-situ monitoring of complex biochemical processes such as

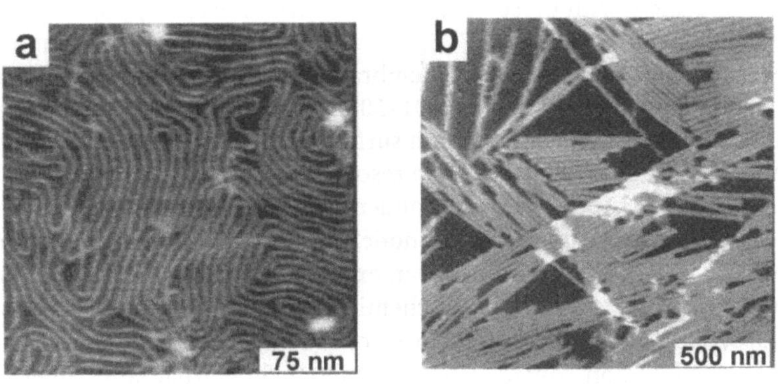

Fig. 54. a A typical image of closely packed pZT plasmid DNA molecules adsorbed onto a cationic lipid membrane of dipalmitoyldimethyl ammoniumylpropane. The image was recorded in 20 mM NaCl [513]. **b** Height image of tobacco mosaic viruses on the mica surface modified with bovine serum albumin (BSA). In contrast, direct adsorption from a solution containing BSA led to disperse adsorption of the TMV rods [517]

transcription and replication [524–526]; (iii) measuring of elastic response of biopolymers to stretching and torsion; (iv) probing specific interactions; and (v) mechanical manipulation of single macromolecules using laser trapping techniques and the scanning force microscope. Some of the recent advances in these fields will be discussed in Sect. 4.3.

4.2
Three Dimensional Polymers

Precise control of the primary structure and the segmental interaction of natural macromolecules enables them to adopt well defined tertiary structures such as supercoils and globules [39]. In contrast, introduction of secondary bonding in synthetic polymers, typically leads to the formation of entangled structures and aggregation. The dense and cascade-type branching provided a useful approach towards three dimensional molecules, which do not interpenetrate each other but interact via their surfaces. Recent synthetic developments include star-like polymers [527], dendrimers [26, 34, 35] and arborescent graft polymers [528, 529]. Polymerisation of substituted macromonomers as well as grafting onto linear chains can yield cylindrically shaped macromolecules such as "hairy rods" [530], worm-like brushes [24, 38, 531, 532] and monodendron-jacketed chains [37, 533, 534]. These colloidal like molecules have been discussed as valuable building units for a future nanotechnology as biochemical sensors [535, 536], molecular containers [537–539], templates for nanolithography [540, 541], energy transfer funnels [542, 543], and polyfunctional initiators and catalysts [539, 544–546]. Hereby, one takes advantage of the narrow molecular weight distribution and the synthetic versatility which enable rigorous control of their size, shape, interior structure and surface functionality.

4.2.1
Molecular Spheres

Interfacial ordering of spherical molecules can be used to prepare a nanoscopic pattern over the entire surface of a planar substrate [540, 547, 548]. When the molecules are adsorbed from solution onto the substrate surface, the ordering process is effected by dewetting events which are driven by the capillary forces emerging during solvent evaporation [518]. This has been studied most intensively for polymer lattices deposited onto a substrate from aqueous suspensions [549–553]. Latex films are of great technological interest since they became widely used for printing pains, lacquers and adhesives. Polymerisation in microemulsions yields ultrafine latex particles ranging from 5 to 50 nm in diameter [554]. Due to the poorly controlled capillary forces and the vigorous Brownian motion, the smallest array of latex spheres achieved so far has a periodicity of 42 nm [551]. Complementary to the densely packed layers, patterned surfaces have been prepared by disperse adsorption of charged latex particles from aqueous solutions [555–558]. Double-layer screening of interparticle electrostatic repulsion was explored to control the surface coverage and hence the interparticle distance. On solid surfaces, the strong adhesion and the irreversible adsorption hinder the thermal mobility of the particles. As a consequence, the equilibration process can be prohibitively long and results only in short range ordering within monolayer domains. Long range ordering in patterned monolayers of latex beads was achieved at the air/water interface due to the higher surface mobility of the water suspended particles [555].

Another promising candidate for preparation of nanoscopic coatings are dendritic molecules such as arborescent graft polymers [528, 529] and dendrimers [26, 34, 35]. The multiple branching yields a large number of chain end functionalities, which distinguishes dendritic macromolecules from conventional star-like polymers [527] and microgels [554]. Moreover, the high branching density prevents overlapping and cause molecular segregation. In contrast to polymer coils, the Euclidean shape, the dense packing of the chain segments, and the topologically defined surface of dendrimers justify considering them as *molecular particles* whose physical properties such as the melt and intrinsic viscosities can approach that of hard spheres [559–562]. The well defined geometry and the low, if any, polydispersity of such molecules favours preparation of nanoscopically ordered films [36]. Figure 55 shows a monolayer structure of the arborescent graft polystyrenes (3rd generation, $M_w \cong 10^7$ g/mol). Except a few defects due to the polydispersity, the monolayer exhibited a fairly ordered structure of the densely packed spheres. Due to the high branching density, the molecules kept their spherical shape, in contrast to the loosely branched polystyrenes which were flattened by adsorption forces. Thus, it is feasible to obtain 2D arrays of nanoscopic spheres with a periodicity from 5 to 100 nm provided that the molecules are monodisperse.

Dendrimers can be regarded as small molecular spheres which are distinguished from other hyperbranched macromolecules by their unprecedented monodispersity. Depending on the generation number, their diameter ranges from 2 to 15 nm [563]. In spite of a certain similarity to macroscopic bodies, dendrimers are too small to be able to neglect the influence of the molecular structure, the chain flexibility and the thermal fluctuations on their physical properties. The ambivalent nature of dendrimers was demonstrated by adsorption and spreading on an interface [315, 319, 564–574]. SFM was used to monitor spreading of a carbosilane dendrimer, which at room temperature represents a complex liquid consisting of globular molecules of about 30 Å in diameter (Fig. 56). Owing to their liquid nature, dendrimer droplets were shown to spread slowly on a flat surface of highly oriented pyrolitic graphite and yield a double layer film. However, owing to their colloidal properties, the monolayer thickness was determined by the globule diameter and the deformability of the particles. Other types of carbosilane dendrimers containing liquid crystal moieties were investigated by SFM regarding their film-formation properties [564]. In this case, the layered structure was promoted by the mesogenic end groups, where the monolayer thickness correlated with the molecular diameter. Self-assembled multilayers were prepared by adsorption of polyamidomine dendrimers using an electrostatic deposition technique [569].

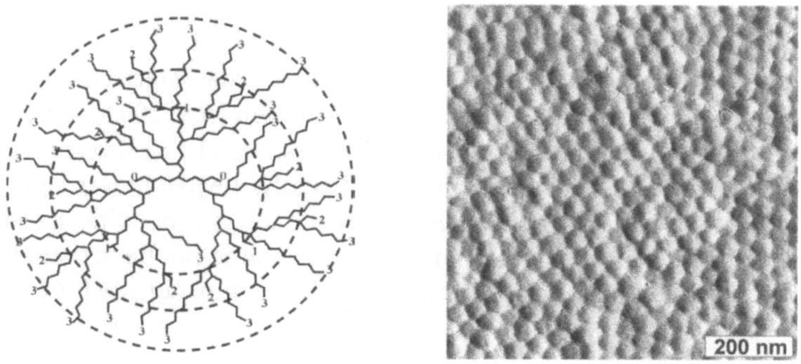

Fig. 55. *left*: Schematic illustration of an arborescent graft polymer as a result of successive grafting reactions. The *dotted circles* and *numbers* correspond to the grafting steps. *right*: SFM images of a monomolecular film of arborescent graft polystyrene S05–3. Each *sphere* corresponds to one molecule having a diameter of 50 nm and a height of 35 nm [36]

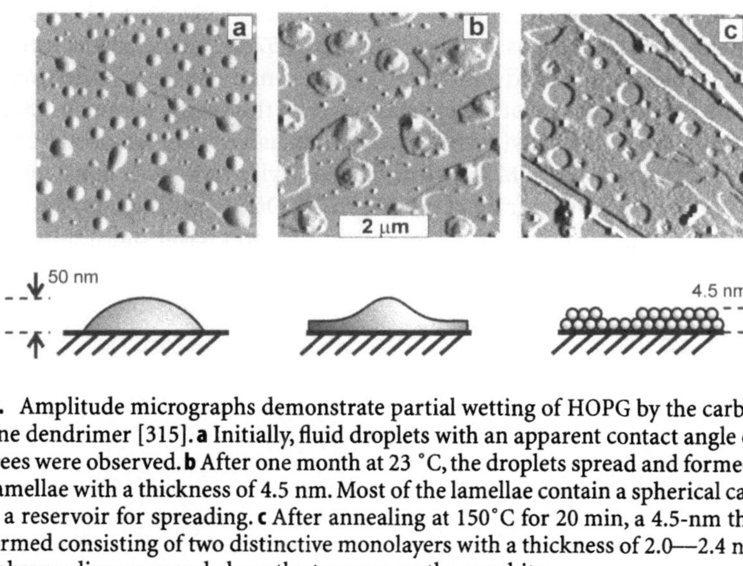

Fig. 56. Amplitude micrographs demonstrate partial wetting of HOPG by the carbosilane-siloxane dendrimer [315]. **a** Initially, fluid droplets with an apparent contact angle of about 8 degrees were observed. **b** After one month at 23 °C, the droplets spread and formed polygonal lamellae with a thickness of 4.5 nm. Most of the lamellae contain a spherical cap which forms a reservoir for spreading. **c** After annealing at 150°C for 20 min, a 4.5-nm thick film was formed consisting of two distinctive monolayers with a thickness of 2.0—2.4 nm. Preferential spreading occurred along the terraces on the graphite

Because of the small size of dendrimers, it is rather difficult to achieve molecular resolution and visualise single molecules by conventional techniques (Sect. 2.3.3). Recently, cryo-TEM has been used to image individual molecules of poly(amidoamine) from generation 5 to G10 [563]. Usually, scanning force microscopy shows some grainy or corrugated surface structure, where the grain size was much greater than the molecular diameter and ascribed to some associates [569–573]. Isolated dendrimer molecules could be prepared by adsorption from dilute solutions using short exposure times [315, 572, 573]. The pictorial resolution of single molecules enabled their characterisation regarding polydispersity [573]. A relatively low size distribution from 100 to 400 nm in diameter for hyperbranched molecules was demonstrated by SFM complementary to light scattering and TEM measurements. Usually, the height was significantly lower (more than two times) as explained by the interaction with the substrate [574]. The anisometric changes of the molecular shape can be prevented by reducing the adhesion forces and increasing the branching density [36].

4.2.2
Worm-like Molecules

Cylindrically shaped macromolecules have been prepared synthetically only rather recently by attaching either linear chains [531, 532, 575, 576] or monodendrons [37, 533, 534] as side groups to a flexible backbone. In the case of cylindrical brushes, the steric repulsion between the densely grafted side chains increases the extension of the backbone [577–579] and leads to a persistence length comparable to the contour length of the main chain [580, 581]. The worm-like conformation of the brushes is illustrated in Fig. 57a. Usually, adsorption of the brush molecules onto a solid substrate yields a densely packed monolayer characterised by a short range orientational order of the brush segments [38, 582]. For example, when the poly(2-vinylpyridine) brushes were adsorbed on mica, they were observed by SFM as semi-flexible cylinders with their backbone aligned parallel to the substrate plane (Fig. 57b). Due to the high grafting density of the side chains, individual molecules in the monolayer did not interpenetrate and formed disordered meanders with hairpins.

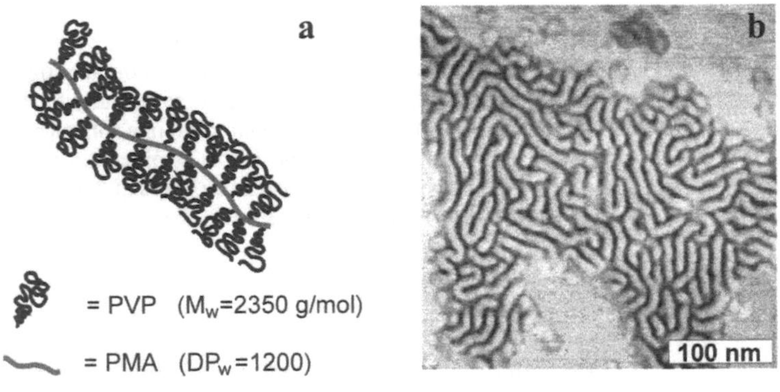

Fig. 57. a Schematic representation of the cylindrical brush consisting of a PMA backbone and PVP side chains (DP_w – weight average degree of polymerisation, M_w – molecular weight of the side chains). **b** Phase micrograph (tapping mode SFM) of PVP brushes deposited from chloroform solution of c=0.1 mg/ml on mica by spin coating. Worm-like molecules organised into a densely packed monolayer [38]

Different methods can be used to prevent molecules from aggregating as they adsorb to the substrate and observe them as single species. This can be done by (i) increasing molecular weight to reduce thermal mobility, (ii) increasing adhesion to the substrate, (iii) modification of the wetting properties of the substrate, and (iv) using rapidly evaporating solvents. Figure 59a shows single molecules of PMMA brushes prepared by adsorbing them from a dilute solution on the polar surface of mica. Visualisation of individual brushes enabled direct investigation of their macroconformations not affected by the neighbouring molecules. By analogy to the pinned micelles in the planar brushes [583–588] axial modulation of density were observed in cylindrical brushes of poly(2-vinylpyridine) [38] and polymethylmethacrylate [589]. PMMA brushes in Fig. 58b display a caterpillar shape, where the periodic variation of the diameter along the main axis were ascribed to the microphase segregation occurring upon increasing the grafting density and decreasing the solvent quality [589]. In this case, the aggregation of the side chains into pinned clusters is driven by the reduction of the unfavourable contacts with the solvent (interfacial energy), which is opposed by the entropic penalty due to the stretching of side-chain legs. Similar effects were observed in the polyelectrolyte chains where electrostatic repulsion provided additional force favouring extension of the backbone [590].

An alternative way to separate molecules is to introduce charges or ionic groups in their chemical structure. For example, poly(2-vinylpyridinium) brushes could be converted into polyelectrolyte brushes by quaternization with benzyl bromide. In aqueous solution, the positively charged brushes adsorb readily onto the negatively charged surface of mica. Figure 59a shows a monolayer structure obtained during in-situ monitoring of the adsorption process by MAC mode SFM [194]. In contrast to the neutral polymers, which aggregate into

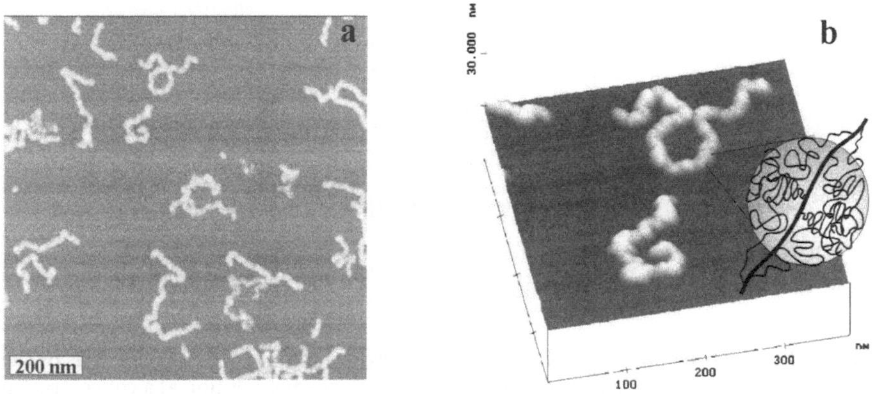

Fig. 58. a A height SFM micrograph of polymethymethacrylate brushes adsorbed on mica from a dilute solution in chloroform [581]. **b** Magnified image of single molecules which exhibit a caterpillar shape. The axial modulations of the diameter have been explained by microphase segregation of the side chains into pinned clusters upon evaporation of the solvent [589]

a densely packed monolayer, charged PVP brushes repel each other, resulting in a sparse coating of worm-like particles. Assuming complete neutralisation of the substrate, the number of adsorbed macromolecules per unit area should be proportional to the charge density of the substrate and reciprocal to the uncompensated charge of the adsorbed species. The degree of coverage was altered between 10 and 50% by changing the ionic strength and the degree of ionisation of the molecules [591]. For example, the coverage of mica with the poly(2-vinylpyridinium) brushes increased two times when 5 mM NaCl was added to the aqueous solution. Figs. 59b-d show a series of images of the poly(2-vinylpyridinium) brushes recorded during adding of NaCl.

Another type of cylindrical macromolecule has been prepared by attaching monodendron substituents as side groups to a linear chain [37, 533, 534]. The tapered shape and a certain amphiphilicity of the monodendrons provides additional factors enforcing the extension of the main chain due to their self-assembly into columns [592]. Depending on the degree of polymerisation and the

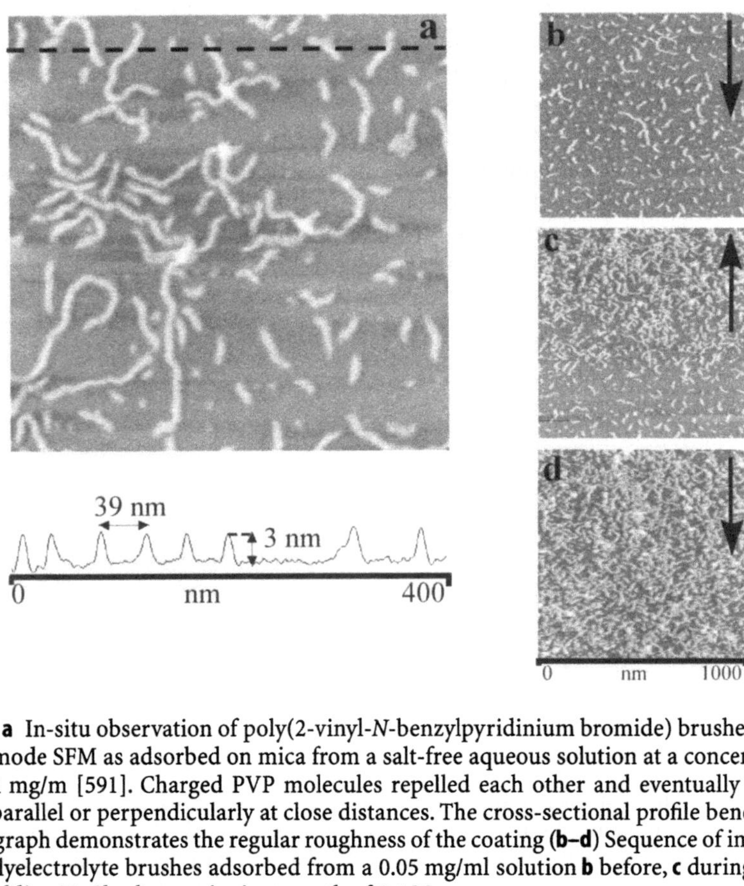

Fig. 59. a In-situ observation of poly(2-vinyl-*N*-benzylpyridinium bromide) brushes by the MAC mode SFM as adsorbed on mica from a salt-free aqueous solution at a concentration of 0.01 mg/m [591]. Charged PVP molecules repelled each other and eventually tend to align parallel or perpendicularly at close distances. The cross-sectional profile beneath the micrograph demonstrates the regular roughness of the coating (**b–d**) Sequence of images of the polyelectrolyte brushes adsorbed from a 0.05 mg/ml solution **b** before, **c** during, and **d** after adding NaCl salt to an ionic strength of 0.1 M

degree of branching of the side groups, the monodendron-jacketed polymers were shown to undergo ordering into a hexagonal columnar or cubic mesophase [593]. On a surface, such molecules were observed by SFM as ca. 5 nm thick threads aligned parallel to the substrate plane [594, 596]. Monodendron-jacketed polystyrenes in Fig. 60a were visualised as single species, when they get adsorbed from solution on HOPG (Fig. 60b). The attained resolution of the chain ends enabled direct measurement of the molecular length. In combination with independently determined molecular weight the chain length per repeat unit could be evaluated as $l_{SFM}=L_w/DP_w$, where DP_w is the degree of polymerisation from static light scattering [594]. The table in Fig. 61a depicts different length values derived from SFM, SEC and SLS experiments. In agreement with the X-ray data [37, 597], the SFM length per monomer unit of polymer 1 (l_{SFM}=1.2 Å) turned out to be almost two times less than the length for a fully extended main chain (l_c=2.5 Å) in an all-*trans* conformation. The axial contraction was attributed to a locally coiled or twisted conformation of the backbone (Fig. 61b). Other polymers of this type with sterically more demanding side groups demonstrated almost complete stretching of the catenated carbon backbone approaching the all-*trans* planar conformation [594, 596].

Additional evidence for the helical conformation of the backbone was found upon adsorption of the monodendron-jacketed polystyrenes on mica which does not exhibit specific interaction with the alkyl tails [596]. Single molecules in Fig. 62a were observed as worm-like cylinders twisted to supercoils called plectoneme (from the Greek meaning "braided string") which represents a way of releasing torsional stress [598, 599]. Due to the intramolecular crossings, the dense monolayer in Fig. 62a resembles a network or felt-like structure. Figure 62b shows a similar conformation observed for biomolecules and model polypeptides [600]. To the best of our knowledge, the supercoil conformation of the monodendron-jacketed polystyrene is one of the first observations of a well defined tertiary structure in synthetic polymers. The plectoneme conformation

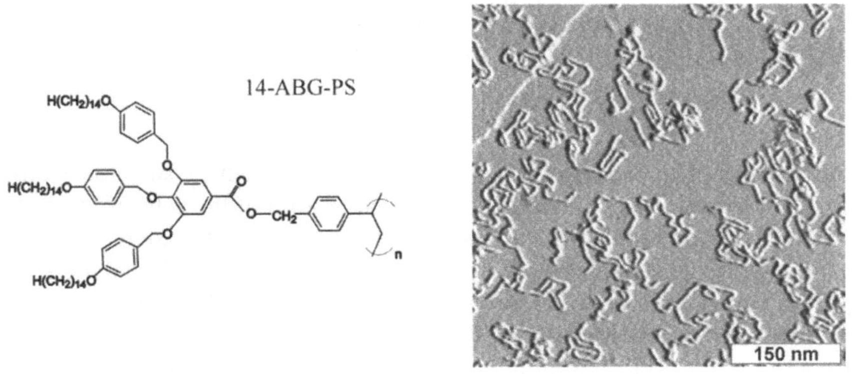

Fig. 60. Chemical structure and SFM micrograph of monodendron-jacketed polystyrene (14-ABG-PS) adsorbed on highly oriented pyrolitic graphite [594]

a

Method	L_w	L_w/L_n	DP_w
SFM	102	1.6	-
SEC-UC	230	2.7	921
SEC-LS	207	2.1	829
SLS	172	-	686

b

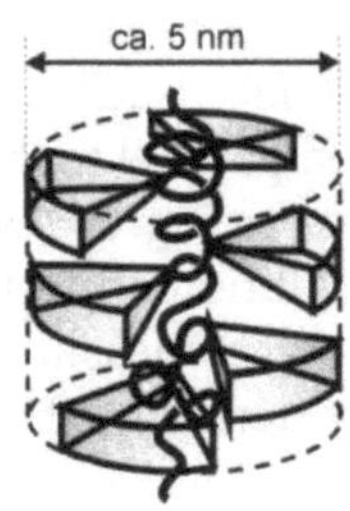

Fig. 61. a The table depicts molecular length (L_w), polydispersity (L_w/L_n), and the degree of polymerisation (DP_w) of the monodendron-jacketed polystyrene 14ABG-PS measured by scanning force microscopy (SFM), size exclusion chromotography with universal calibration (SEC-US) and light scattering detector (SEC-LS), and static light scattering (SLS). **b** A random helix conformation of a linear chain with monodendron side groups was proposed to explain the axial contraction of the monodendron-jacketed linear chain [594]

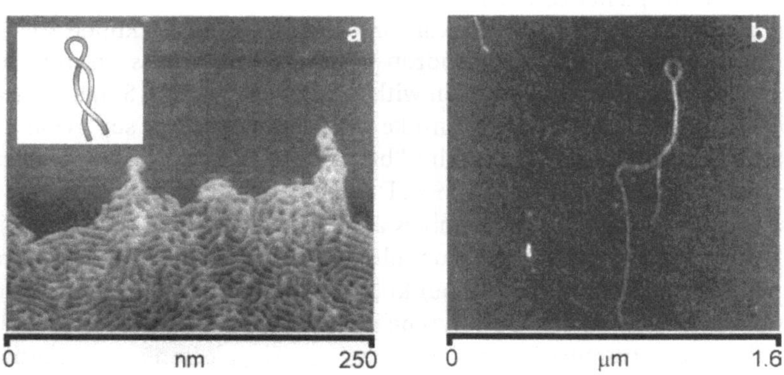

Fig. 62. High resolution height SFM-micrographs of **a** 14-ABG-PS on mica [595] and **b** twisted ribbon structure of polypetide β-sheets [599]. Protruding molecules at the film edge (marked by *arrows*) clearly show the plectoneme type conformation caused by the backward folding of the torsionally stressed molecules. Insert in (**a**) depicts a plectoneme supercoil

could be caused by underwinding or overwinding of a helical backbone from its equilibrium state [601]. Obviously, quick evaporation of the solvent resulted in a residual torsion in the molecule as it contracted in itself. Unlike macroconformations of biomolecules, where the tertiary structures are often stabilised by specific interactions between side groups, the supercoil of the monodendron-jacketed polymers is metastable. Eventually, annealing offered a path for the stress relaxation and allowed the structural defects to heal [596].

A subsequent question, which has been explicitly addressed, regards the interaction of the brush like molecules with the flat solid substrate [594, 596]. The height of the cylindrical brushes [38, 582] as well as monodendron-jacketed polymers was determined to be considerably smaller than their lateral diameter. This flattening indicates strong interaction of the side chains to the substrate. Besides changes in the molecular shape, strong adsorption can cause new types of macroconformations and even ordering of the brush-like macromolecules. In Fig. 63a, polystyrene, where each styrene unit is substituted by a 3,4,5 tribenzyloxybenzoate with *one* tetradecyloxy substituent in the *para*-position of the benzyl groups, display a peculiar conformation consisting of straight segments with bends of a characteristic angle of 60 ° and 120 °. In contrast, the same polymer did not show any particular order on mica (Fig. 63b). Only a short-range orientational order of densely packed cylinders was observed. The ordering of the macromolecules was explained by the epitaxial adsorption of the alkyl tails of the monodendron side groups on HOPG. Formation of highly ordered monolayers due to preferential orientation of the hydrocarbon chains with respect to the graphite lattice was reported for linear alkanes and cycloalkanes [602, 603], as well as for comb-like molecules with alkyl substituents such as olygothiophenes [604]. In this case, the repeat length l_e=2.54 Å of a $-(CH_2-CH_2)-$ sequence in an all-*trans* planar zigzag conformation matches the crystallographic spacing a=2.46 Å of the graphite surface. In the case of monodendron-substituted polymers, the configurational accessibility of the tails to the adsorption sites depended on the degree of branching of the side groups. For polymers, where each dendron is substituted by nine alkyl substituents, resulting in a rather dense coat around the backbone, the epitaxial adsorption was significantly hindered due to steric repulsion between the side chains and greater stiffness of the backbone.

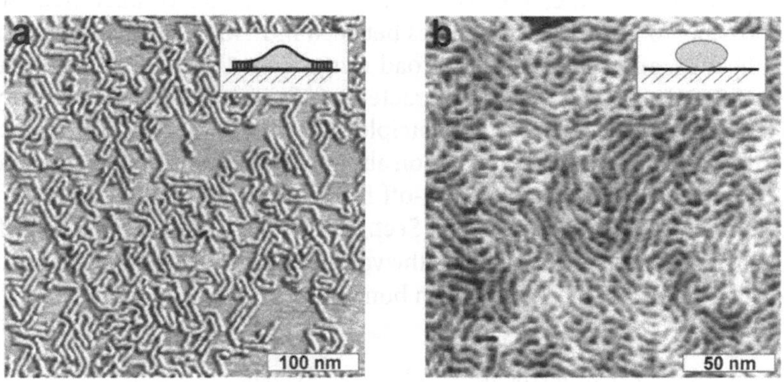

Fig. 63. SFM-micrographs of the monolayers on **a** HOPG and **b** mica prepared by spincasting of 14-ABG-PS solutions in cyclopentane (c=0.1 mg/ml). Individual molecules in (**a**) aligned parallel to the substrate and bent at characteristic angles of 60 ° and 120 °. This indicates epitaxial adsorption of the alkyl tails according to the three fold symmetry of graphite. The worm-like molecules in (**b**) wound around each other and resulted in a felt-like structure [596]. The inserts illustrate specific interaction of the side chains with HOPG (**a**) and their absence on mica (**b**)

In conclusion, the pictorial resolution of single macromolecules by SFM provided a new analytical technique for polymer characterisation regarding molecular dimensions and macroconformations. The resolution has been achieved due to the well defined shape and the strong molecular segregation of the hyperbranched polymers. Another factor, which facilitates the observation individual molecules, is adhesion to the substrate. However, one should take great care in interpreting their macroconformation which can be greatly affected by the interaction between the side chains and the substrate surface. The combination of the interfacial interactions with the intramolecular steric repulsion acting along the backbone can be a powerful tool to organise nanoscopic cylinders on a surface.

4.3 Probing Specific Interactions and Elasticity of Individual Molecules

In addition to the covalent bonds, nature has developed a variety of multiple nonocovalent bonds that demonstrate a high degree of spatial and directional specificity. These molecular-recognition interactions play a key role in assembly and regulation in living organisms. During the last decade, different techniques such as micropipette [605], microfibres [606], optical or magnetic tweezers [607–610], surface force apparatus [611], and scanning force microscopy [612, 613] have been employed to measure molecular forces and elasticity of single molecules with a sensitivity down to 0.1 pN (Table). These measurements appeared to be complementary to other stretching experiments, using hydrodynamic flows [614] and electric fields [615], as well as theoretical models of a stressed polymer chain [616–618].

Compared to the other methods, SFM probes smaller contact areas which make possible to investigate the forces between individual objects such as molecules, ligands, and cells, within a broad range from 1 pN to 100 nN. Different modes of SFM have been used to characterise adhesion between chemically defined surfaces (see Sect. 2.2.2). In principle, chemical force microscopy is capable of providing qualitative information about both the cohesion energy and the work of adhesion by measuring pull-off force or contact area between similar and dissimilar surfaces, respectively (Sect. 3.3). However, still it remains difficult to extract from these measurements the value of a single bond force. Some estimations have been done for hydrogen bonds [611].

Performance data of SFM, micropipette and optical tweezers

parameters	SFM	micropipette	optical tweezers
force constant, pN/nm	10–100	0.1–5	10^{-3}–10^{-2}
vertical resolution nm	0.1	2–5	1–5
loading rate, pN/s	10^4–10^6	0.1–10^5	1–10
force resolution, pN	5–20	0.2–0.5	0.01–0.1

It has been relatively more successful to measure adhesion forces between large biomolecules such as biotin-streptavidin [222, 619] antibody and antigen, and complementary strands of DNA [380, 620]. In these experiments, glass microspheres functionalized with the molecules were used as a probe to measure rupture forces. Although the measurements were reproducible, the interpretation appeared to be difficult because of the inherent complexity of this type of interaction. For example, the streptavidin-biotin interaction involves at least seven hydrogen bonds and a similar number of hydrophobic interactions [621]. This makes it rather problematic to attribute the observed forces to a single interaction act and to estimate the force value from the unknown potential [619]. Moreover, the bond strength was shown to increase from about 5 pN to 170 pN with increasing the loading rate from 10^{-1} to 10^{5} pN/s [605]. Therefore, one of the strongest linkages in biology can be stronger or weaker depending on how fast they are loaded. Even more complicated is the interpretation of interaction forces between single stranded DNA where one has to distinguish between adhesive forces resulted from complementary, non-complementary and non-specific interactions of oligonucleotide sequences [620].

Many peculiar functions of biomolecules are based on their unique ability of specific folding of their polypeptide chains. Qualitatively, this property was investigated by chemical and thermal denaturation experiments. Scanning force microscopy has been employed to probe specific forces between binding pairs e.g. in polysaccharides like native dextran and xantan [622, 623]. In addition to the intramolecular assembling, some of the proteins have to withstand mechanical forces to maintain a certain structure in dynamic processes. For example, titin is the giant sarcomeric protein (M=2.5 MDa) which controls the elasticity of striated muscles. The elasticity of titin results from the reversible unfolding of the linear array of ca. 300 immunoglobulin (Ig) and fibronectin type III domains [624, 625]. Because of the unknown potential of such complicated molecules, the unfolding forces have to be measured directly by either SFM [626] or optical tweezers [627]. In SFM measurements, the force curves showed a distinctive set of sawtooth peaks with a periodicity of 25–28 nm, and peak force varying from 150 to 300 pN. Since the periodicity corresponded to the contour length 31 nm of an unravelled Ig domain, the profile was ascribed to the successive unravelling of individual domains of a single titin molecule. Somewhat different observations were made using optical tweezers. Instead of the sawtooth profile, the stretching of titin molecules using the optical tweezers led to a continuous increase of the force up to 25–30 pN [627]. Some of the molecules could be extended far beyond the 1 μm contour length of native titin [628]. The discrepancy has been explained based on kinetic arguments due to difference in the pulling rates [476]. The latter was considerably lower for optical tweezers, i.e. 650 nm/s, compared to 0.5 mm/s used in the SFM experiments. In both case, the force-distance dependence could be described by the Kratky-Porod model for worm-like chains [495–499]. In the stretching experiments, titin displayed relaxation behaviour and hysteresis effects. Refolding occurred only when the molecule was allowed to relax completely. The unfolding force, it was suggested, depends logarithmically on the extension speed [629]. Experimentally, stiffer cantilevers have been proposed for unfolding experiments, whereas softer cantilevers might be of advantage to follow the molecular refolding [476].

Complementary to stretching of proteins, different single-molecule probes have been used to study supercoiling of DNA. These experiments are expected to provide direct evidence which would link the biological activity of DNA and its tertiary structure. So far, the fantastic ability of DNA molecules (in total ~3 meters! long divided into 46 chromosomes) to pack into the micrometer sized nucleus of human cells requires better understanding [601]. It is clear that the random coil conformation would not fit inside the nucleus. Recently, a series of elegant experiments have been performed to stretch an individual DNA strand under torsionally relaxed as well as constrained conditions [380, 601, 610, 630, 631]. These experiments have been inspired by the work of Smith et al, where single DNA molecules were chemically attached by one end to a glass surface and their other end to a magnetic bead driven by an external magnetic field [607]. The force-distance curves in Fig. 64a exhibit a S-type shape where small and steady increase in the force was followed by a steep increase and an extended plateau [601]. At the beginning, a low contractile force (<5 pN) corresponds to the entropic elasticity. The forces rise rapidly up to ca. 70 pN where the contour length is approached. The plateau was ascribed to the unwinding and splitting of DNA molecules. The S-shape of the curve can be approximated by an interpolation formula which describes the extension x of a semiflexible chain [498] with contour length L_0 in response to a stretching force F [499, 617]:

$$\frac{FP}{k_B T} = \frac{1}{4}\left(1 - \frac{x}{L_0}\right)^{-2} - \frac{1}{4} + \frac{x}{L_0}$$

Where k_B is the Boltzmann constant and T is the absolute temperature. The equation describes the entropic elasticity of the chain with a persistence length P assuming that the DNA molecule is inextensible. Apparently, this mechanical denaturation, or the so-called "enthalpic elasticity", exhibited strong dependence on the buffer conditions [610]. Figure 64b shows the effect of the ionic strength on the persistence length of λ-bacteriophage DNA. The flexibility of the molecules was increasing upon adding salt which lead to P values as low as 45–50 nm for monovalent ions (Fig. 64b).

Recently, successful experiments have been carried to visualise single chains of synthetic polymers and probe their elasticity. Usually, polymer chains immobilised on a rigid polymer surface tend to aggregate into globular clusters [632, 633]. A simple procedure has been proposed to avoid chain agglomeration and obtain single polymer coils [634]. A mixture of thiol terminated PS chains and alkanethiols was chemisorbed onto a gold surface from a toluene solution. This resulted in isolated polymer chains randomly distributed in a self-assembled monolayer of alkylthiols. Thus prepared samples were used to investigate elasticity of individual polymer chains [635].

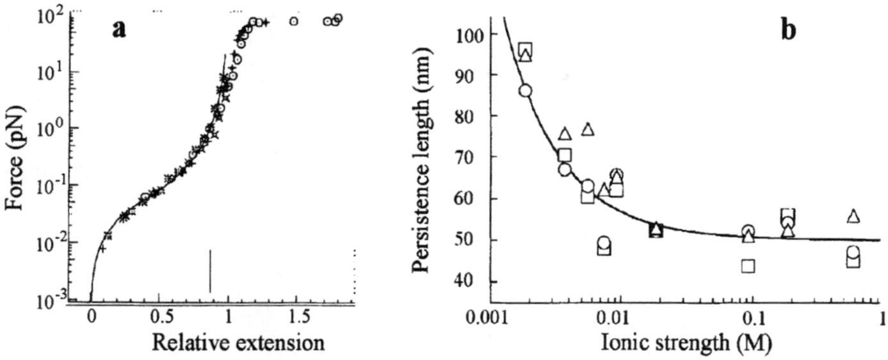

Fig. 64. a Force versus extension curves measured on beads bound by a single λ-DNA. At low stretching forces (0.02<F<2 pN) the DNA molecule in 10 mM phosphate buffer behaves like a worm-like chain with a persistence of 53 nm (continuum line). At higher forces, the molecule extends beyond its contour length from 1.1 to 1.8 times [601]. **b** Dependence of the persistence length (P) on monovalent (Na+) ionic strength. *Points* correspond to the three WLC fits: ❑ inextensible WLC, ○ strong-stretching limit, Δ extensible WLC [610]

4.4 Manipulation of Single Atoms and Small Molecules

In the preceding section we demonstrated the efficiency of SFM in visualisation and probing of organic molecules. So far, the method has been less efficient in manipulation and moving of individual molecules on a surface. In this respect, scanning tunnelling microscopy has been shown to be a more powerful tool. Here we have to admire the elegant experiments of Eigler et al. at IBM Almaden and Gimzewski et al. at IBM Rüschlikon [435, 436, 636–640]. They have made remarkable progress in the placement of atoms in a desired pattern. Different types of molecules could be repositioned laterally by pushing them with the STM tip, or they could be transferred from a surface to the tip and the back again. Among many fundamental problems, the inertia law does not hold exactly since the diffusion barrier height rather than the mass controls the relation between the applied force and the resulting acceleration. In these experiments, one has to control the interaction between the molecule and the surface, and trigger the interaction between the molecule and the tip. From the latter aspect, STM is more efficient compared to scanning force microscopy. A series of images in Fig. 65 were obtained from very inspiring experiments where Bohringer et al. have manipulated vacancies on a silver surface and the researchers at the IBM Research Station in Rüschlikon have built an abacus of C_{60} beads [437].

These experiments demonstrate a peculiar crossover of two conceptually different approaches in nanofabrication, i.e. the top-down approach using lithographic techniques and the bottom-up approach using self-assembly of elementary building blocks. From one side, constructing things one-atom-at-a-time

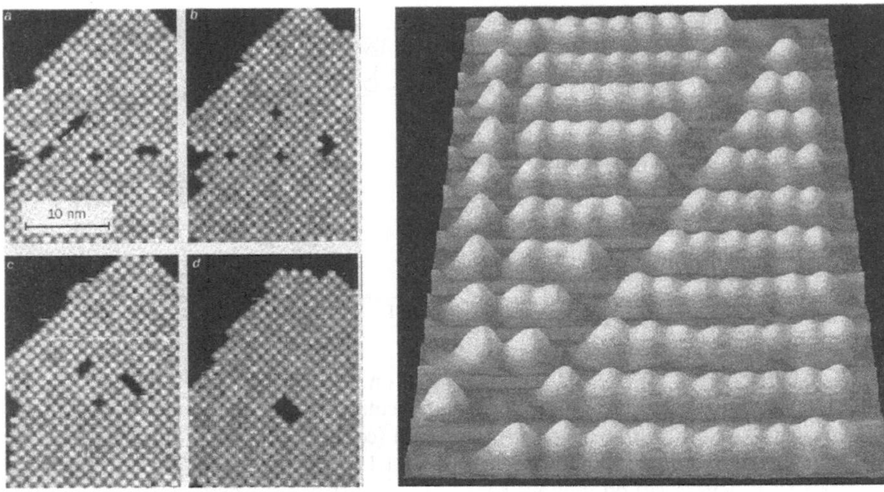

Fig. 65. (*left*) Vacancies on a silver surface can be manipulated by using an STM tip to move a neighbouring molecule into the gap. Here Bohringer et al. (University of Lausanne) show how the six vacancies in (**a**) can be repositioned into a rectangular cluster (**d**). (*right*) The beads in this abacus are C_{60} molecules less than 1 nm in diameter positioned on a silicon surface. They can be pushed back and forth with an STM tip in a controlled fashion to count from 1 to 10

follow the general idea of lithography to construct a desired circuit and not an arbitrary pattern. On the other hand, it represents an assembly of desired molecules into a thermodynamically permitted structure. This so-called "molecular Lego" approach represents one of the greatest challenges in fabrication of molecular devices and pushed the frontier of nanotechnology towards single-molecule electronics.

5 Summary

Over the last 5 years, scanning force microscopy has demonstrated a series of remarkable developments from a simple imaging technique towards a technique which can probe interfacial properties and manipulate structures on the nanoscale. Concerning *imaging*, the major improvements have occurred in the ability of SFM to analyse surface structures (i) non-destructively and (ii) in a physically relevant environment. This is of particular importance for biological samples which now can be imaged in-situ retaining their native structure and biological activity during durable experiments. Apart from routine measurements on the micrometer scale, the resolution of the surface features became reliable on the nanometer scale, but still remain rather ambiguous on the atomic scale. Only special techniques which enable near to contact scanning in a vacuum provided reproducible images of atomic defects indicating that the achieved resolution is true. Meanwhile, SFM has been able to visualise single macromolecules with sizes of around 5 nm and thus has become a new analytical method for the characterisation of polymers regarding their size, flexibility and macroconformation.

Concerning *properties*, SFM has become a unique technique in probing local adhesion, friction and elastic response of various materials. This is based on the ability to measure forces as small as picoNewtons and probe areas well below 100 nm. The peculiar sensitivity of the force probe to different types of static and dynamic interactions provides a great number of contrast mechanisms which can map the surface structure regarding the chemical composition and physical properties. However, in most SFM measurements the interpretation of the surface maps remain to be very intricate, mostly because of the concurrent contribution of different forces into the net force. The progress in this field relies on new developments in technique which would allow to measure the properties like stiffness, adhesion, friction and viscosity, separately.

Finally, scanning probe microscopy techniques, including SFM, have been established as new *nanofabrication* methods featuring the size scale below 1 nm. In a highly sophisticated way, they combine the manual control of lithographic techniques with thermodynamically controlled assembly of single molecules. The ability of SPM to fabricate on the nanometer scale and furthermore construct things with atoms has strongly pushed the frontier in nanotechnology towards molecular devices and exposed a wealth of unexplored problems and *terra incognito* in various disciplines of the fundamental science.

Acknowledgements. I am deeply grateful to Martin Möller, Manfred Schmidt and Yuli Godovskij for comments on the text and the contents of the article. I am also indebted to B. Anczykowski, G. Bar, N.A. Burnham, C. Bustamante, L. Leibler, R. Lüthi, S.N. Magonov, G. Reiter, C. Rivetti, U. Steiner, T. Russell as well as my colleagues at the University of Ulm U. Beginn, P. Eibeck, S. Hild, S. Mößmer, A. Mourran, S. Prokhorova, K. Weishaupt for providing useful references and original figures. A great part of the reviewed work, in particular on surface organisation of single macromolecules (Sect. 4), was financially supported by the Deutsche Forschungsgemeinschaft.

References

1. Gelbart WM, Ben-Shaul A, Roux D (eds) (1994) Micelles, membranes, microemulsions, and monolayers. Springer, Berlin Heidelberg New York
2. Goodwin JW, Buscall R (eds) (1995) Colloidal polymer particles. Academic Press, London
3. Jönsson B, Lindman B, Holmberg K, Kronberg B (1998) Surfactants and polymers in aqueous solution. Wiley, Chichester
4. Ikada Y, Uyama Y (1993) Lubricating polymer purfaces. Technomic, Basel
5. Wool RP (1995) Polymer interfaces:structure and strength. Hanser, Munich
6. Berg JC (ed) (1993) Wettability. Marcel Dekker, New York
7. Scheirs J (ed) (1997) Modern fluoropolymers:high performance polymers for diverse applications. Wiley Series in Polymer Science. Wiley, Chichester
8. Singhi R, Kumar A, Lopez GP, Stephanopoulos GN, Wang DIC, Whitesides GM, Ingber DE (1994) Science 264:696

9. Aksay IA, Trau M, Manne S, Honma I, Yao N, Zhou L, Fenter P, Eisenberg PM, Gruner SM (1996) Science 273:892
10. Russell TP (1996) Curr Opin Colloid and Interface Sci 1:107
11. Park M, Harrison C, Chaikin PM, Register RA, Adamson DH (1997) Science 276:1401
12. Möller M, Spatz JP (1997) Curr Opin Colloid Interface Sci 2:177
13. Muthukumar M, Ober CK, Thomas EL (1997) Science 277:1225
14. Ober CK, Wegner G (1997) Adv Mater 9:17
15. Ruokolainen J, Mäkinen R, Torkkeli M, Mäkelä T, Serimaa R, ten Brinke G, Ikkala O (1998) 280:557
16. Böltau M, Walheim S, Mlynek J, Krausch G, Steiner U (1998) Nature 391:877
17. Rabe JP (1998) Curr Opin Colloid Interf Sci 3:27
18. Ulrich R, Chesne AD, Templin M, Wiesner U (1999) Adv Mater 11:141
19. Mann S (1996) Biomimetic materials chemistry. VCH, Weinheim
20. Ball P (1997) Made to measure. New materials for 21st century. Prinecton University Press, New Jersey
21. Fendler JH (1998) Nanoparticles and nanostructured films:preparation, characterization and applications. Wiley-VCH, Weinheim
22. Ito H, Reichmanis E, Nalamasu O, Ueno T (eds) (1998) Micro- and nanopatterning polymers. ACS Symposium Series 706. American Chemical Society, Washington DC
23. Timp G (ed) (1999) Nanotechnology. Springer, Berlin Heidelberg New York
24. Yamashita Y (ed) (1993) Chemistry and industry of macromonomers. Hüthig & Wepf Verlag, Basel Heidelberg New York
25. Brintzinger HH, Fischer D, Mülhaupt R, Rieger B, Waymouth RM (1995) Angew Chem Int Ed Engl 34:1143
26. Newkome GR, Moorefield CN, Vögtle F (eds) (1996) Dendritic molecules. concept, synthesis, perspectives. VCH, Weinheim
27. Matyjaszewski K (ed) (1998) Controlled radical polymerization. ACS Symposium Series 685. American Chemical Society, Washington DC
28. Fredrickson GH, Bates FS (1996) Ann Rev Mater Sci 26:501
29. Jung K, Abetz V, Stadler R (1996) Macromolecules 29:1076
30. Antonietti M, Förster S (1998) Adv Mater 10:195
31. Hajduk DA, Harper PG, Gruner SM, Honeker LC, Kim G, Thomas EL, Fetters LJ (1994) Macromolecules 27:4063
32. Radzilowski LH, Carragher BO, Stupp SI (1997) Macromolecules 30:2110
33. Liu G (1998) Curr Opin in Colloid and Interface Sci 3:200
34. Dvornic PR, Tomalia DA (1996) Curr Opin Colloid Interf Sci 1:221
35. Fréchet JMJ, Hawker CJ, Gitsov I, Leon JW (1996) JMS – Pure Appl Chem A33:1399
36. Sheiko SS, Gauthier M, Möller M (1997) Macromolecules 30:2343
37. Percec V, Ahn C-H, Cho W-D, Jamieson AM, Kim J, Leman T, Schmidt M, Gerle M, Möller M, Prokhorova SA, Sheiko SS, Cheng SZD, Zhang A, Ungar G, Yeardley DJP (1998) J Am Chem Soc 120:8619
38. Dziezok P, Sheiko SS, Fischer K, Schmidt M, Möller M (1997) Angew Chemie Int Ed Engl 109:2812
39. Alberts B, Bray D, Lewis J, Raff M, Roberts K, Watson JD (1994) Molecular biology of the cell. 3rd Ed. Garlan Publishing Inc. New York London
40. Alivisatos AP, Johnsson KP, Peng X, Wilson TE, Loweth CJ, Bruchez Jr. MP, Schulz PG (1996) Nature 382:609
41. Mirkin CA, Letsinger RL, Mucic RC, Storhoff JJ (1996) Nature 382:607
42. Wetzig K, Schulze D (1995) In situ scanning electron microscopy in materials research. Akademie Verlag, Germany
43. Brune D, Hellborg R, Whitlow HJ, Hunderi O (1997) Surface characterization. Wiley-VCH, Weinheim
44. Amelinckx S, van Dyck D, van Landuyt J, van Tendeloo G (eds) (1997) Handbook of microscopy. VCH, Weinheim

45. Woodward AE (1989) Atlas of polymer morphology. Hanser, Munich Vienna New York
46. Hooke R (1665) Some physiological descriptions of minute bodies made by magnifying glasses. Martyn and Allestry, London
47. Abbe E (1873) Arch Mikrosk Anat 9:413
48. Wilson T (ed) (1990) Confocal Microscopy. Academic Press, London
49. Ruska E (1933) Z Phys 83:684
50. Binnig G, Rohrer H, Gerber Ch, Weibel E (1982) Phys Rev Lett 49:57
51. Bonnell DA (ed) (1993) Scanning tunneling microscopy and spectroscopy:theory, techniques, and applications. VCH, New York Weinheim
52. Marti O (1993) SXM:an introduction. In: STM and SFM in biology. Academic Press, London
53. Sarid D (1994) Scanning force microscopy. Oxford University Press, New York
54. Güntherodt HJ, Weisendanger R (eds) Scanning tunneling microscopy III. Springer Series in Surface Sciences 29, 2nd edn. Springer, Berlin Heidelberg New York
55. Fillard JP (1996) Near field optics and nanoscopy. World Scientific, Singapore
56. Paesler M, Moyer PJ (1996) Near field optics:theory, instrumentation and applications. Wiley, New York
57. Colton RJ, Engel A, Frommer JE, Gaub HE, Gewirth AA, Guckenberger R, Rabe J, Heckl WM, Parkinson B (eds) (1998) Procedures in scanning probe microscopies. Wiley, Chichester
58. Magonov SN, Whangbo M-H (1996) Surface analysis with STM and AFM. VCH, Weinheim
59. Frommer J (1992) Angew Chem Int Ed Engl 104:1298
60. MaMaster TJ, Carr HJ, Miles MJ, Cairno P, Morris VJ (1991) Macromolecules 24:1428
61. Rabe JP, Buchholz S (1991) Phys Rev Lett 66:2096
62. Williamson RL, Miles MJ (1996) SPIE Proceeding 2535:82
63. Sandoghdar V, Wegscheider S, Krausch G, Mlynek J (1997) J Appl Phys 81:2499
64. Binnig G, Quate CF, Gerber Ch. (1986) Phys Rev Lett 56:930
65. Magonov SN, Reineker DH (1997) Annu Rev Mater Sci 27:175
66. Tsukruk VV (1997) Rubber Chemistry and Technology 70:430
67. Ratner BD, Tsukruk VV (eds) (1998) Scanning probe microscopy of polymers. ACS Symposium Series 694. American Chemical Society, Washington DC
68. Jandt KD (1998) Mater Sci Eng R21:221
69. Kajiyama T, Tanka K, Ge S-R, Takahara AT (1996) Progress in Surface Science 52:1
70. Burnham NA, Colton RJ (1993) Force microscopy. In: Bonnell DA (ed) Scanning Tunneling Microscopy and Spectroscopy: Theory, Techniques, and Applications. VCH, New York Weinheim
71. Lindsay SM (1993) Biological applications of the scanning probe microscope. In: Bonnell DA (ed) Scanning tunneling microscopy and spectroscopy:theory, techniques, and applications. VCH, New York Weinheim
72. Meyer E, Heinzelmann H (1996) Scanning force microscopy. In: Güntherodt HJ, Weisendanger R (eds) Scanning tunneling microscopy II. Springer series in surface sciences 28, 2nd edn. Springer, Berlin Heidelberg New York, p 99
73. Israelachvili JN, Tabor D (1972) Proc R Soc Lond A 331:19
74. Klein J (1980) Nature 288:248
75. Williamson JPB (1967) Proc Inst Mech Eng London 182:21
76. Clayton TE, Scire FE, Baker SM, Jensen SW (1982) Wear 83:1
77. Israelachvili J (1992) Intermolecular & Surface Forces. 2d ed. Academic Press, San Diego
78. Adamson AW (1995) Physical chemistry of surfaces, 5th edn. Wiley, New York London
79. Hertz H (1881) J Reine Angew Math 92:156
80. Johnson KL, Kendall K, Roberts AD (1971) Proc R Soc London A 324:301
81. Derjaguin BV, Muller VM, Toporov YP (1975) J Coll Interface Sci 53:314
82. Maugis D (1992) J Coll Interf Sci 150:243
83. Dugdale DS (1960) J Mech Phys Solids 8:100
84. Magonov SN, Sheiko SS, Debliek RAC, Möller (1993) Macromolecules 26:1380

85. Vancso GJ (1996) Polym Prepr (Am Chem Soc, Div Polym Chem) 37:550
86. Meyer G, Amer NM (1988) Appl Phys Lett 53:2400
87. Alexander S, Hellemans L, Marti O, Schneier J, Elings V, Hasma PK, Longmire M, Gurley J (1989) J Appl Phys 65:164
88. Mate CM, McClelland GM, Erlandsson R, Chiang S (1987) Phys Rev Lett 59:1942
89. Meyer G, Amer NM (1990) Appl Phys Lett 57:2089
90. Marti O, Colchero J, Mlynek J (1990) Nanotechnology 1:141
91. Martin Y, Williams CC, Wickramasinghe HK (1989) J Appl Phys 61:4723
92. Senden TJ, Drummond CJ, Kékicheff P (1994) Langmuir 10:358
93. Manne S, Cleveland JP, Gaub HE, Stucky GD, Hansma PK (1994) Langmuir 10:4409
94. Sarid D, Elings V (1991) J Vac Sci Technol B 9:431
95. Meyer E (1992) Prog Surf Sci 41:3
96. Hölscher H, Schwarz DU, Zwörner O, Wiesendanger R (1998) Phys Rev B 57:2477
97. Jarvis SP, Yamada H, Yamamotio S-I, Tokumoto H, Pethica JB (1996) Nature 384:247
98. Bammerlin M, Lüthi R, Meyer E, Baratoff A, Lü J, Guggisberg M, Loppacher C, Gerber C, Güntherodt HJ (1998) Appl Phys A 66:293
99. Hu J, Xiao X-D, Salmeron M (1995) Appl Phys Lett 67:476
100. Kikukawa A, Hosaka S, Honda, Imura R (1995) Rev Sci Instrum 66:101
101. Zhong Q, Innis D, Kjoller K, Elings VB (1993) Surf Sci Lett 290:L688
102. Whangbo M-H, Bar G, Brandsch R (1998) Surf Sci 411:L794
103. Rao SS (1995) Mecanical Vibrations. Addison-Wesley, New York
104. Bar G, Brandsch R, Whangbo M-H (1998) Surf Sci 411:L802
105. Kühle A, Sorensen AH, Bohr J (1997) J Appl Phys 81:6562
106. Bachelot R, Gleyzes P, Boccara AC (1998) Probe Microscopy 1:89
107. Marth M, Maier D, Honerkamp J, Brandsch R, Bar G (1999) J Appl Phys 85:1
108. Radmacher M, Tillmann RW, Gaub HE (1993) Biophys J 64:735
109. Bar G, Thomann Y, Brandsch R, Cantow H-J, Whangbo M-H (1997) Langmuir 13:3807
110. Magonov SN, Elings V, Whangbo M-H (1997) Surf Sci 375:L385
111. Noy A, Sanders CH, Vezenov DV, Wong SS, Lieber CM (1998) Langmuir 14:1508
112. Krüger D, Anczykowski B, Fuchs H (1997) Annalen der Physik 6:341
113. Anczykowski B, Cleveland JP, Krüger D, Elings V, Fuchs H (1998) Appl Phys A 66:S885
114. Digital Instruments Inc. Santa Barbara, CA, USA
115. Tribology Issues and Opportunities in MEMS (1998) Bhushan B (ed) Kluwer, Dordrecht
116. Lantz MA, O'Shea SJ, Welland ME, Johnson KL (1997) Phys Rev B 55:10776
117. Kendall K (1994) Science 263:1720
118. Carpick RW, Ogletree DF, Salmeron M (1997) Appl Phys Lett 70:1548
119. Lantz MA, O'Shea SJ, Hoole ACF, Welland ME (1997) Appl Phys Lett 70:970
120. Johnson KL (1997) Proc R Soc London Ser A 453:163
121. Ferry JD (1980) Viscoelastic Properties of Polymers. Wiley, New York
122. Burnham NA, Kulik AJ, Gremaud G, Gallo P-J, Oulevey F (1996) J Vac Sci Technol B 14:794
123. Maivald P, Butt HJ, Gould SAC, Prater CB, Drake B, Gurley JA, Elings VB, Hansma PK (1991) Nanotechnology 2:103
124. Bar G, Rubin S, Parikh AN, Swanson BI, Zawodzinski TA, Whangbo M-H (1997) Langmuir 13:373
125. Kajiyama T, Takata K, Ohki I, Ge S-R, Yoon J-S, Takahara (1994) Macromolecules 27:7932
126. Nysten B, Legras R, Costa J-L (1995) 78:5953
127. Friedenberg MC, Mate CM (1996) Langmuir 12:6138
128. Galuska AA, Poulter RR, McElrath KO (1997) Surface and Interface Analysis 25:418
129. Radmacher M, Cleveland JP, Fritz M, Hansma HG, Hansma PK (1994) Biophys J 66:2159

130. O´Shea SJ, Welland ME, Pethica JB (1994) Chem Phys Lett 223:336
131. Florin E-L, Radmacher M, Fleck B, Gaub HE (1994) Rev Sci Instrum 65:639
132. Mazeran P-E, Loubet J-L (1997) Tribol Lett 3:125
133. Tanaka K, Taura A, Ge SR, Takahara A, Kajiyama T (1996) Macromolecules 29:3040
134. Burnham NA, Gremaud G, Kulik AJ, Gallo P-J, Oulevey F (1996) J Vac Sci Technol B 14:1308
135. Overney RM, Leta DP, Pictroski CF, Rafailovich MH, Liu Y, Quinn J, Sokolov J, Eisenberg A, Overney G (1996) Phys Rev Lett 76:1272
136. Wahl KJ, Stepnowski SV, Unertl WN (1998) Tribol Lett 5:103
137. Overney RM, Takano H, Fujihira M (1994) Europhys Lett 26:443
138. Oulevey F, Gremaud G, Sémoroz, Kulik AJ, Burnham NA, Dupas E, Gourdon D (1998) Rev Sci Instrum 69:2085
139. Hammische A, Reading M, Pollock HM, Song M, Hourston DJ (1996) Rev Sci Instrum 67:4268
140. Winkler RG, Spatz JP, Sheiko S, Möller M, Reineker P, Marti O (1996) Phys Rev 54:8908
141. Sarid D, Ruskell TG, Workman RK, Chen D (1996) J Vac Sci Technol B 14:864
142. Tamayo J, García R (1997) Appl Phys Lett 71:2394
143. Burnham NA, Behrend OP, Oulevey F, Gremaud G, Gallo P-J, Gourdon D, Dupas E, Kulik AJ, Pollock HM, Briggs GAD (1997) Nanotechnology 8:67
144. Cleveland JP, Anczykowski B, Schmid AE, Elings VB (1998) Appl Phys Lett 72:2613
145. Grütter P, Liu Y, LeBlanc P, Düring U (1997) Appl Phys Lett 71:279
146. Anczykowski B, Gotsmann B, Fuchs H, Cleveland JP, Elings VB (1999) Appl Surf Sci (in press)
147. Bar G, Brandsch R, Whangbo M-H (1999) Surf Sci 422:L192
148. Cooke PM (1998) Anal Chem 70:385R
149. Overney RM, Meyer E, Frommer, Brodbeck D, Lüthi R, Howald L, Güntherodt H-J, Fujihira M, Takano H, Gotoh Y (1992) Nature 359:133
150. Overney RM, Meyer E, Frommer, Güntherodt H-J (1994) 10:1281
151. Frisbie CD, Roznay LF, Noy A, Wrighton MS, Lieber CM (1994) Science 265:2071
152. Akari S, Horn D, Keller W, Schrepp W (1995) Adv Mater 7:549
153. Berger CE, van der Werf KO, Kooyman RP, de Grooth BG, Greeve J (1995) Langmuir 11:4188
154. Frommer JE (1996) Thin Solid Films 273:112
155. Noy A, Vezenov DV, Lieber CM (1997) Annu Rev Mater 27:381
156. Van der Werf KO, Putman CAJ, Groth BG, Greve J (1994) Apll Phys Lett 65:1195
157. Rosa A, Weilandt E, Hild S, Marti O (1997) Meas Sci and Technol 8:1
158. Miyatani T, Horii M, Rosa A, Fujihira M, Marti O. (1997) Appl Phys Lett 71:2632
159. Hild S, Krotil U, Marti O (1998) In: Bhushan B (ed) Tribology Issues and Opprtunities in MEMS. Kluwer, Dordrecht, p 247
160. Knoll B, Keilmann F (1999) Nature 399:134
161. Hammiche A, Pollock HM, Song M, Hourston DJ (1996) J Meas Sci Technol 7:142
162. Nakagawa Y, Schäfer R, Güntherodt H-J (1998) Apll Phys Lett 73:2296
163. Berger R, Lang HP, Gerber Ch, Gimzewski JK, Fabian JH, Scandella L, Meyer E, Güntherodt H-J (1998) Chem Phys Lett 294:363
164. Poler CC, Zimmermann RM, Cox EC Langmuir (1995) 11:2689
165. Hörber JKH, Häberle W, Ohnesorge F, Binnig G, Lieibich HG, Czerney CP, Mahnel H, Mayr A (1992) Scanning Microsc 6:919
166. Strausser YE, Heaton MG (1994) Am Lab 26:20
167. Kamusewitz H, Keller M, Paul D (1995) Thin Solid Films 264:184
168. Beaglehole D, Christenson HK (1992) J Phys Chem 96:3395
169. Hu J, Xiao X-D, Ogletree DF, Salmeron M (1995) Science 268:267
170. Yang G, Vesenka JP, Bustamante CJ (1996) Scanning 18:344
171. Colchero J, Storch A, Luna M, Herrero JG, Baró AM (1998) Langmuir 14:2230
172. Giessibl FJ Gerber Ch, Binnig G (1995) Science 267:68
173. Giessibl FJ, Gerber Ch, Binnig G (1992) J Vac Sci Technol B 9:984

174. Neubauer G, Cohen SR, Horn D, Mate CM, McClelland GM (1990) Rev Sci Instr 61:2296
175. Germann GJ, Cohen SR, Neubauer G, McClelland GM, Seki H (1993) J Appl Phys 73:163
176. Meyer G, Amer NM (1990) Appl Phys Lett 57:2089
177. Weisenhorn AL, Hansma PK, Albrecht TR, Quate CF (1989) Appl Phys Lett 54:2651
178. Ducker WA, Senden TJ, Pashley RM (1992) Langmuir 8:1831
179. Senden TJ, Drummond CJ, Kékichef P (1994) Langmuir 10:358
180. Meyer E, Howald L, Overney RM, Heinzelmann H, Frommer J, Güntherodt H-J, Wagner T, Schier H, Roth S (1991) Nature 349:398
181. Butt H-J, Downing KH, Hansma PK (1990) Biophys J 58:1473
182. Cleveland JP, Schäffer TE, Hansma PK (1995) Phys Rev B 52:R8692
183. Han W, Lindsay SM (1998) Appl Phys Lett 72:1656
184. Ohnesorge F, Binnig G (1993) Science 260:1451
185. Putman CAJ, van der Werf KO, de Grooth BG, van Hulst NF, Greve J (1994) Appl Phys Lett 64:2454
186. Hansma PK, Cleveland JP, Radmacher M, Walters DA, Hilner PE, Bezanilla M, Fritz M, Vie D, Hansma HJ, Prater CB, Massie J, Fukunada L, Gurley J, Elings V (1994) Appl Phys Lett 64:1738
187. Lantz MA, O'Shea SJ, Welland ME (1994) Appl Phys Lett 65:409
188. Anselmetti D, Dreier M, Lüthi R, Richmond T, Meyer E, Frommer J, Güntherodt H-J (1994) J Vac Sci Technol B 12:1500
189. Putman CAJ, van der Werf KO, de Grooth BG, van Hulst NF, Greve J (1994) Biophys J 67:2454
190. Fritz M, Radmacher M, Cleveland JP, Allersma MW, Stewart RJ, Gieselmann R, Janmey P, Schmidt CF, Hansma PK (1995) Langmuir 11:3529
191. Lee GU, Chrisey LA, Colton RJ (1994) Science 266:771
192. Walters DA, Cleveland JP, Thomson NH, Hansma PK, Wendman MA, Gurley G, Elings V (1996) Rev Sci Instrum 67:3583
193. Elmer FJ, Dreier M (1997) J Appl Phys 81:7709
194. Han W, Lindsay SM, Jing T (1996) Appl Phys Lett 69:4111
195. O'Shea SJ, Welland ME, Pethica JB (1994) Chem Phys Lett 223:336
196. Florin E-L, Radmacher M, Fleck B, Gaub HE (1994) Rev Sci Instrum 65:639
197. Albrecht TR, Akamine S, Carver TE, Quate CF (1990) J Vac Sci Technol A 8:3386
198. Wolter O, Bayer T, Greschner J (1991) J Vac Sci Technol B 9:1353
199. Buser RA, Bruger J, de Rooij NF (1992) Ultramicroscopy 42:1476
200. Pechmann R, Köhler JM, Fritzsche W, Schaper A, Jovin TM (1994) Rev Sci Instrum 65:3702
201. Keller DJ, Chih-Chung C (1992) Surface Science 268:333
202. Keller D, Deputy D, Alduino A, Luo K (1992) Ultramicroscopy 42–44:1481
203. Fischer BP, Wie MS, Chou SY (1993) J Vac Sci Technol B 11:2570
204. Tortonese M (1997) In: Cohen SH, Lightbody ML (eds) Atomic Force Microscopy / Scanning Tunneling Microscopy. Vol 2, Plenum, New York, p 147
205. Linnenmann R, Gotszalk T, Rangelow IW, Dumania P, Oesterschulze EJ (1996) J Vac Sci Technol B 14:856
206. Akama Y, Nishimura E, Sakai A, Murakami H (1990) J Vac Sci Technol A 8:429
207. Hansma HG, Vesenka J, Siegerist C, Kelderman G, Morrett H, Sinsheimer RL, Elings V, Bustamante C, Hansma PK (1992) 256:1180
208. Walters DA, Hampton D, Drake B, Hansma HG, Hansma PK (1994) Appl Phys Lett 65:787
209. Vasile MJ, Grigg DA, Griffith JE, Fitzgerald EA, Russell PE (1991) J Vac Sci Technol B 9:3569
210. Ximen H, Russell PE (1992) Ultramicroscopy 42–44:1526
211. Boisen A, Hansen O, Bouwstra S (1996) J Micromech Microeng 6:58
212. Fotino M (1992) Appl Phys Lett 60:2935
213. Dai H, Hafner JH, Rinzler AG, Colbert DT, Smalley RE (1996) Nature 384:147
214. Rutland MW, Senden TJ (1993) Langmuir 9:412

215. Whitesides GM, Laibinis PE (1990) Langmuir 6:87
216. Dubois LH, Nuzzo RG (1992) Annu Rev Phys Chem 43:437
217. Frisbie CD, Noy A, Rozsnyai LF, Lieber CM, Wrighton MS (1995) J Am Chem Soc 117:7943
218. Thomas RC, Tangyunyoung P, Houston JE, Michalske TA, Crooks RM (1994) J Phys Chem 98:4493
219. Nakagawa T, Ogawa K, Kurumizawa T, Ozaki S (1993) Jpn J Appl Phys 32:L294
220. Kumar A, Whitesides GM (1993) Appl Phys Lett 14:63
221. Barrat A, Silberzahn P, Bourdieu L, Chatenay D (1992) Europhys Lett 20:633
222. Florin E-L, Moy VT, Gaub HE (1994) Science 264:415
223. Tsao Y-H, Evans DF, Wennerström H (1993) Science 262:547
224. Grütter P, Zimmermann-Edling W, Brodbeck D D (1992) Appl Phys Lett 60:2741
225. Sheiko SS, Möller M, Reuvekamp EMCM, Zandbergen HW (1994) Ultramicroscopy 53:371
226. Schwartz DU, Haefke H, Reimann P, Güntherodt H-J (1994) J Microsc 173:183
227. Reiss G, Brückl H, Vancea J, Lecheler R, Hastreiter E (1991) J Appl Phys 70:523
228. Paik SM, Kim S, Schuller IK (1991) Phys Rev Lett B 44:3272
229. Melmed AJ (1991) J Vac Sci Technol B 9:601
230. Garnaes J, Kragh F, Morch KA, Tholen AR (1990) J Vac Sci Technol A 8:441
231. Grigg DA, Russell PE, Griffith JE, Vasile MJ, Fitzgerald EA (1992) Ultramicroscopy 42–44:1616
232. Edenfeld KM, Jarausch KF, Stark TJ, Griffis DP, Russel PE (1994) J Vac Sci Technol B 12:3571
233. Atamny F, Baiker A (1995) Surf Sci 323:L314
234. Vesenka J, Miller R, Henderson E (1994) Rev Sci Instrum 65:2249
235. Xu S, Arnsdorf MF (1994) J Microsc 173:199
236. Sheiko SS, Möller M, Reuvekamp EMCM, Zandbergen HW (1993) Phys Rev B 48:5675
237. Carpick RW, Agrait N, Ogletree DF, Salmeron M (1996) J Vac Sci Technol B 14:1289
238. Gracias DH, Somorjai GA (1998) Macromolecules 31:1269
239. Heuberger M, Dietler G, Schlapbach (1996) J Vac Sci Technol B 14:1250
240. Weihs TP, Nawaz Z, Jarvis SP, Pethica JB (1991) Appl Phys Lett 59:3536
241. Pethica JB, Oliver WC (1987) Physica Scripta T19:61
242. Jarvis SP, Pethica JB (1995) In: Güntherodt H-J, Anselmetti D, Meyer E (eds) Forces in Scanning Probe Methods. NATO ASI Series, Ser E, vol 286. Kluwer, Dordrecht, p 105
243. Pethica JB, Sutton AP (1988) J Vac Sci Technol A 6:2490
244. Colton RJ, Baker SM, Driscoll RJ, Youngquist MG, Baldeschwieler JD, Kaiser WJ (1988) J Vac Sci Technol A 6:349
245. Landman U, Luedke WD, Nitzan A (1989) Surf Sci Lett 10:L177
246. Overney RM, Takano H, Fujihira M, Paulus W, Ringsdorf H (1994) Phys Rev Lett 72:3546
247. Garnaes J, Schwarz DK, Viswanathan R, Zasadzinski JAN (1992) Nature 255:64
248. Viswanathan R, Madsen LL, Schwartz DK, Zasadzinski JA (1995) Science 269:51
249. Bourdieu L, Ronsin O, Chatenay D (1993) Science 259:798
250. Butt H-J, Seifert E, Bamberg E (1993) J Phys Chem 97:7316
251. Bustamante C, Keller D (1995) Physics Today December:32
252. Hashimoto T, Tanaka H, Hasegawa H (1990) Macromolecules 23:4378
253. Matsen MW, Bates FS (1997) J Chem Phys 106:2436
254. Leibler L (1980) Macromolecules 13:1602
255. Semenov AN (1985) Sov Phys JETP 61:733
256. Bates FS, Fredrickson GH (1990) Ann Rev Phys Chem 41:525
257. Fredrickson GH (1987) Macromolecules 20:2535
258. Turner MS, Rubinstein M, Marques CM (1994) Macromolecules 27:4986
259. Anastasiadis SH, Russell TP, Satija SK, Majkrzak CF (1989) Phys Rev Lett 62:1852
260. Collin B, Chatenay D, Coulon G, Ausserre D, Gallot Y (1992) Macromolecules 25:1621

261. Ausserre D, Raghunathan VA, Maaloum M (1993) J Phys II (France) 3:1485
262. Schull KR (1993) Macromolecules 26:2346
263. Liu Y, Zhao W, Zheng X, King A, Singh A, Rafailovich MH, Sokolov J, Dai KH, Kramer EJ, Schwarz SA, Gebizlioglu O, Sinha SK (1994) Macromolecules 27:4000
264. Leclére Ph, Lazzaroni R, Brédas JL, Yu JM, Dubois PH, Jérome R (1996) Langmuir 12:4317
265. Magonov SN, Cleveland J, Elings V, Denley D, Whangbo M-H (1997) Surface Sci 389:201
266. McLean R, Sauer BB (1997)Macromolecules 30:8314
267. Pickering JP, Vancso GJ (1998) Polym Bull 40:549
268. Gattiglia E, Turturro A, Ricci D, Bonfiglio A (1995) Macromol Rapid Comm 16:919
269. Van Dijk MA, van der Berg R (1995) Macromolecules 28:6773
270. Kellog GJ, Walton DG, Mayes AM, Lambooy P, Russell TP, Gallagher PD, Satija SK (1996) Phys Rev Lett76:2503
271. Walton DG, Kellogg, Mayes AM, Lambooy P, Russell TP (1994) Macromolecules 27:6225
272. Lambooy P, Russell TP, Kellogg GJ, Mayes AM, Gallagher PD, Satija SK (1994) Phys Rev Lett 72:2899
273. Kikuchi M, Binder K (1994) J Chem Phys 101:3367
274. Picket GT, Balazs AC (1997) Macromolecules 30:3097
275. Reiter G Castelein G, Hoerner P, Riess G, Blumen A, Sommer J-U (1999) submitted to Science
276. Spatz JP, Möller M, Noeske M, Behm J, Pietralla M (1997) Macromolecules 30:3874
277. Spatz JP, Eibeck P, Mößmer S, Möller, Herzog T, Ziemann P (1998) Adv Mater 10:849
278. Koutsos V, van der Vegte E, Pelletier E, Stamouli A, Hadziionaou G (1997) Langmuir 4719
279. Potemkin II, Kramarenko EYu, Khokhlov AR, Winkler RG, Reineker P, Eibeck P, Spatz J, Möller M (1999) Langmuir (in press)
280. Zhulina EB, Singh C, Balazs AC (1996) Macromolecules 29:6338
281. Russell TP (1995) communication at the workshop on September 18–21, Konstantz
282. Heier J, Kramer EJ, Walheim S, Krausch G (1997) Macromolecules 30:6610
283. Morkved TL, Witzius P, Jaeger HM, Grier DG, Witten TA (1994) Appl Phys Lett 64:422
284. Morkved TL, LU M, Urbas AM, Ehricks EE, Jaeger HM, Mansky P, Russell TP (1996) Science 273:931
285. Tuzar Z, Kratochvil P (1993) In Surface and Colloid Science. (ed) Matijevic E. Plenum, New York
286. Chu B (1995) Langmuir 11:414
287. Förster S, Zisenis M, Wenz E, Antonietti M (1996) J Chem Phys 104:9956
288. Yu Y, Zhang L, Eisenberg A (1998) Macromolecules 31:1144
289. Spatz JP, Sheiko SS, Möller M (1996) Macromolecules 29:3220
290. Meiners JC, Quintel-Ritzi A, Mlynek J, Elbs H, Krausch G (1997) Macromolecules 30:4946
291. Spatz JP, Mößmer S, Möller M (1996) Chem Europ J 1:1552
292. Hempenius MA, Langeveld BMW, van Haare JAEH, Janssen RAJ, Sheiko SS, Spatz JP, Möller M, Meijer EW (1998) J Am Chem Soc 120:2798
293. Spatz JP, Mößmer S, Möller, Herzog T, Plettl A, Ziemann P (1998) J Luminescence 76–77:168
294. Spatz JP, Herzog T, Mößmer S, Ziemann P, Möller (1999) Adv Mater 11:149
295. Stocker W, Rabe JP, Abetz W, Stadler R (1996) Macromolecules 29:7502
296. Magerle R, Elbs H, Fukunage K, Sauer G, Stadler R, Krausch G (1999) Macromolecules 32:1204
297. de Gennes P-G (1985) Rev Mod Phys 57:827
298. Cazabat A-M (1987) Contemp Phys 28:347

299. Leger L, Joanny JF (1992) Rep Prog Phys 55:431
300. Zhao W, Rafailovich MH, Sokolov J, Fetters LJ, Plano R, Sanyal Mk, Sinha K (1993) Phys Rev Lett 70:1453
301. Leibler L, Ajdari A, Mourran A, Coulon G, Chatenay D (1994) In: Teramoto A, Kobayashi M, Norisuje T (eds) Ordering in Macromolecular Systems. Springer-Verlag, Berlin Heidelberg
302. Tanaka K, Takahara A, Kajiyama T (1998) Macromolecules 31:863
303. Robbins MO, Andelmann D, Joanny J-F (1991) Phys Rev A 43:4344
304. Israelachvili JN, McGuiggan P, Homola AM (1988) Science 240:189
305. Brochard F, de Gennes P-G (1984) J Phys Lett 45:L597
306. Bruinsma R (1990) Macromolecules 23:275
307. Silberzan P,Leger L (1992) Macromolecules 25:1267
308. Valignat MP, Oshanin G, Villette S, Cazabat AM, Moreau M (1998) Phys Rev Lett 80:5377
309. Brochard F, di Meglio JM, Quéré D, de Gennes P-G (1991) Langmuir 7:335
310. Reiter G, Sharma A, Casoli A, David M-O, Khanna R, Auroy P (1999) Langmuir 15:2551
311. Joanny J-F, de Gennes P-G (1986) J Colloid Interface Sci 111:94
312. Gaydos J, Neumann AW (1987) J Colloid Interface Sci 120:76
313. Rieutord F, Salmeron M (1998) J Phys Chem B 102:3941
314. Tamayo J, García R (1996) Langmuir 12:4430
315. Sheiko SS, Eckert G, Ignat'eva G, Muzafarov AM, Spickermann J, Räder HJ, Möller M (1996) Macromol Rapid Comm 17:283
316. Pompe T, Fery A, Herminghaus S (1998) Langmuir 14:2585
317. Boruvka L, Gaydos J, Neumann AW (1990) Colloids Surf 43:307
318. Drelich J (1996) Colloids & Surf A 116:43
319. Sheiko SS, Muzafarov AM, Winkler RG, Getmanova EV, Eckert G, Reineker P (1997) Langmuir 13:4172
320. Zisman WA (1963) Industrial and Engineering Chemistry 55:18
321. Brochard F, Daillant J (1990) Can J Phys 68:1084
322. Reiter G (1993) Langmuir 9:1344
323. Sheiko SS, Lermann E, Möller M (1996) Langmuir 12:4015
324. Sheiko SS, Krupers M, Slangen P-J, Möller M (1998) Polymer Prepr. 39(2):956
325. Stange TG, Evans DF, Hendrickson WA (1997) Langmuir 13:4459
326. Krausch G, Hipp M, Böltau M, Marti O, Mlynek J (1995) Macromolecules 28:260
327. Kressler J, Wang C, Kammer HW (1997) Langmuir 13:4407
328. Jacobs K, Herminghaus S, Mecke KR (1998) Langmuir 14:965
329. Segalman R, Green PF (1999) Macromolecules 32:801
330. Ausserré D, Brochard-Wyart B, de Gennes P-G (1995) C.R.Acad. Sci. (Paris) Série II 320:131
331. Feng Yi, Karim A, Weiss RA, Douglas JF, Han CC (1998) Macromolecules 31:484
332. Liu Y, Rafailovich MH, Sokolov J, Schwarz SA, Zhong X, Eisenberg A, Kramer EJ, Sauer BB, Satija S (1994) Phys Rev Lett 73:440
333. Long D, Ajdari A, Leibler L (1996) Langmuir 12:1675
334. Reiter G, Auroy P, Auvray L (1996) Macromolecules 29:2150
335. Shull KR (1996) Macromolecules 29:8487
336. Kerle T, Cohen SR, Klein J (1997) Langmuir 13:6360
337. Kerle T, Yerushalmi-Rozen, Klein J (1998) Macromolecules 31:422
338. Andelmann D, Joanny J-F, Robbins MO (1988) Europhys Lett 7:731
339. Tidswell IM, Rabedeau TA, Pershan PS, Kosowsky SD (1991) Phys Rev Lett 66:2108
340. Li Z, Tolan M, Höhr T, Kharas D, Qu S, Sokolov J, Rafailovich MH, Lorenz H, Kotthaus JP, Wang J, Sinha SK, Gibaud A (1998) Macromolecules 31:1915
341. Krausch G (1995) Mater Sci Eng Rep R14:1
342. Kumar SK, Russell TP (1991) Macromolecules 24:3816

343. Zhao X, Zhao W, Sokolov J, Rafailovich MH, Schwarz SA, Wilkens BJ, Jones RAL, Kramer EJ (1991) Macromolecules 24:5591
344. Jones RAL, Norton LJ, Kramer EJ, Bates FS, Wiltzius P (1991) Phys Rev Lett 66:1326
345. Krausch G, Dai C-A, Kramer EJ, Marko JF, Bates FS (1993) Macromolecules 26:5566
346. Steiner U, Klein J, Eiser E, Budkowski A, Fetters LJ (1992) Science 258:1126
347. Straub W, Bruder F, Brenn R, Krausch G, Bielefeldt H, Kirsch A, Marti O, Mlynek J, Marko JF (1995) Europhys Lett 29:353
348. Jandt KD, Heier J, Bates FS, Kramer EJ (1996) Langmuir 12:3716
349. Tanaka K, Yoon J-S, Takahara A, Kajiyama T (1995) Macromolecules 28:934
350. Müller-Boschbaum P, O'Neill SA, Affrossman S, Stamm M (1998) Macromolecules 31:5003
351. Steiner U, Klein J, Fetters LJ (1994) Phys Rev Lett 72:1498
352. Walheim S, Böltau M, Mlynek J, Krausch G, Steiner U (1997) Macromolecules 30:4995
353. Sung L, Karim A, Douglas JF, Han CC (1996) Phys Rev Lett 76:4368
354. Keblinski P, Kumar SK, Maritan A, Koplik J, Banavar JR (1996) Phys Rev Lett 76:1106
355. Ermi BD, Karim A, Douglas JF (1998) J Polym Sci, Polym Phys 36:191
356. Karim A, Slawecki TM, Kumar SK, Douglas JF, Han CC, Russell TP, Liu Y, Rafailovich M (1998) Macromolecules 31:857
357. Mayes AM, Kumar SK (1997) MRS Bulletin 22:43
358. Jones RAL, Kramer EJ (1993) Polymer 34:115
359. Jalbert C, Koberstein JT, Hariharan A, Kumar SK (1997) Macromolecules 30:4481
360. Lopez GP, Biebuyck HA, Frisbie CD, Whitesides GM (1993) Science 260:647
361. Gau H, Herminghaus S, Lenz P, Lipowsky R (1999) Science 283:46
362. Krausch G, Kramer EJ, Rafailovich MH, Sokolov J (1994) Appl Phys Lett 64:2655
363. Ermi BD, Nisato G, Douglas JF, Rogers JA, Karim A (1998) Phys Rev Lett 81:3900
364. Balazs AC, Huang K, McElwain P, Brady JE (1991) Macromolecules 24:714
365. de Gennes P-G (1980) Macromolecules 13:1069
366. Milner ST (1991) Science 251:905
367. Halperin A, Tirrell M, Lodge TP (1992) Adv Polym Sci 100:31
368. Tirrell M (1997) Curr Opin Colloid Interface Sci 2:70
369. Luckham PF (1996) Curr Opin Colloid Interface Sci 1:39
370. Gast AP (1996) Langmuir 12:4060
371. Klein J, Kumacheva E, Perahia D, Mahalu D, Warburg S (1994) Faraday Discuss 98:173
372. Claesson PM, Blomberg E, Frobert JC, Nylander T, Arbebrant T (1995) Adv Colloid Interface Sci 57:161
373. Kilbey SM, Bates FS, Tirrell M, Yoshizawa H, Hill R, Israelachvili J (1995) Macromolecules 28:5626
374. Kuhl T, Guo YQ, Alderfer JL, Berman AD, Leckband D, Israelachvili J, Hui SW (1996) Langmuir 12:3003
375. Pelletier E, Lapique F, Montfort JP (1996) Colloid Surf 112:43
376. Abe T, Kurihara K, Higashi N, Niwa M (1995) J Phys Chem 99:1820
377. Dahlgren M, Hollenberg H, Claesson P (1995) Langmuir 11:4480
378. Parsegian VA, Evans EA (1996) Curr Opin Colloid Interface Sci 1:53
379. Claesson PM, Edeth T, Bergeron V, Rutland MW (1996) Adv Colloid Interface Sci 67:119
380. Smith SB, Cui Y, Bustamante C (1996) Science 271:795
381. Manne S, Gaub HE (1997) Curr Opin Colloid Interface Sci 2:145
382. Butt HJ, Jaschke M, Ducker W (1995) Bioelectrochem Bioenerget 38:191
383. Sader JE, Larson I, Mulvaney P, White LR (1995) Rev Sci Instr 66:3789
384. Cleveland JP, Manne S, Bocek D, Hansma PK (1993) Rev Sci Instr 64:403
385. Ducker WA, Senden TJ, Pashley RM (1991) Nature 353:239
386. Wood J, Sharma R (1995) Langmuir 11:4797
387. Chaudhury MK, Whitesides GM (1992) Science 255:1230
388. Mate CM (1995) IBM J Res Develop 39:617

389. Noy A, Frisbie CD, Rozsnyai LF, Wrighton MS, Lieber CM (1995) J Am Chem Soc 117:7943
390. Liu Y, Evans DF, Song Q, Graigner DW (1996) Langmuir 12:1235
391. Thomas RC, Houston JE, Crooks RM, Kim T, Michalske TA (1995) J Am Chem Soc 117:3830
392. Braithwaite GJC, Luckham PF (1997) J Chem Soc, Faraday Trans 93:1409
393. Milling A, Mulvaney P, Larson I (1996) J Colloid Interface Sci 180:460
394. Raiteri R, Grattarola M, Butt HJ (1996) J Phys Chem 100:16700
395. Cai LL, Peanasky J, Granick S (1996) Trends Polym Sci 4:57
396. Biggs S (1996) J Chem Soc Faraday Trans 92:2783
397. Zhulina E, Walker GC, Balazs AC (1998) Langmuir 14:4615
398. Grest GS, Lacasse M-D, Murat M (1997) MRS Bull 22:27
399. Roters A, Schimmel M, Rühe J, Johannsmann D (1998) Langmuir 15:3999
400. Zhang J, Uchida E, Uyama Y, Ikada Y (1997) J Colloid Interface Sci 188:431
401. Hudson JE, Abruna HD (1996) J Am Chem Soc 118:6303
402. Hillier AC, Kim S, Bard AJ (1996) J Phys Chem 100:18808
403. Milling AJ, Biggs S (1995) J Colloid Interface Sci 170:604
404. Milling AJ (1996) J Phys Chem 100:8986
405. O'Shea SJ, Welland ME, Rayment T (1992) Appl Phys Lett 60:2356
406. Klein J, Kumacheva E. (1995) Science 269:816
407. Horn RG, Hirz SJ, Hadziioannou GH, Frank CW, Catala JM (1989) J Chem Phys 90:6767
408. Ducker WA, XU Z, Israelachvili JN (1994) Langmuir 10:3279
409. Yang ACM, Terris B, Kunz M (1991) Macromolecules 24:6800
410. Sheiko SS, Kunz M, Möller M (1994) Polym Sci USSR 35:1609
411. Hild S, Gutmannsbauer W, Lüthi R, Fuhrmann J, Güntherodt H-J (1996) J Polym Sci:Part B Polym Phys 34:1953
412. Carpick RW, Salmeron M (1997) Chem Rev 97:1163
413. Overney RM (1995) Trends Polym Sci 3:359
414. Nyffenegger RM, Penner RM (1997) Chem Rev 97:1195
415. Chizhik SA, Huang Z, Gorbunov VV, Myshkin NK, Tsukruk VV (1998) Langmuir 14:2606
416. Boschung E, Heuberger M, Dietler G (1994) Appl Phys Lett 64:3566
417. Howard AJ, Rye RR, Houston JE (1996) J Appl Phys 79:1885
418. Radmacher M, Fritz M, Kacher CM, Cleveland JP, Hansma PK (1996) Biophys J 70:556
419. Aimé JP, Gauthier S (1998) in Scanning Probe Microscopy of Polymers. ACS Symposium Series 694. Ratner BD, Tsukruk VV (eds) American Chemical Society, Washington DC
420. Nie H-Y, Motomatsu M, Mizutani W, Tokumoto H (1996) Thin Solid Films 273:143
421. Yamamoto S, Yamada H (1997) Langmuir 13:4861
422. Krim J (1996) Sci Am 275:74
423. Reiley TC, Fan L-S, Mamin HJ (1995) Microelectron Eng 27:495
424. Bhushan B, Kulkarni AV (1996) Thin Solid Films 278:49
425. Bhushan B, Koinkar VN (1994) Appl Phys Lett 64:1653
426. Haugstad G, Gladfelter WL, Jones RR (1996) J Vac Sci Technol A 14:1864
427. Werts MPL, van der Vegte EW, Grayer V, Esselink E, Tsitsilianis C, Hadziioannou G (1998) Adv Mater 10:452
428. Koinkar VN, Bhushan B (1996) J Appl Phys 79:8071
429. Mueller A, Kowalewski T, Wooley KL (1998) Macromolecules 31:776
430. Smith HI, Gaighead HG (1990) Physics Today Feb:24
431. Quate CF (1994) Surf Sci 299:980
432. Mamin HJ, Rugar D (1992) Appl Phys Lett 61:1003
433. Avouris P (1995) Acc Chem Res 28:95
434. Kern DP (1994) Microelectron Eng 23:41
435. Gimzewski JK, Joachim C, Schlittler RR, Langlais V, Tang H, Johannsen I (1998) Science 281:531

436. Gimzewski J (1998) Physics World 11:29
437. Kolb DM, Ullmann R, Will T (1997) Science 275:1097
438. Kranz C, Ludwig M, Gaub HE, Schuhmann (1995) 7:38
439. Marrian CRK, Dobisz EA, Colton RJ (1991) J Vac Sci Technol B 9:1367
440. Albrecht TR, Quate CF (1987) J Appl Phys 62:2599
441. Snow ES, Juan WH, Pang SW, Campbell PM (1995) Appl Phys Lett 66:1729
442. Teuschler T, Mahr K, Miyazaki S, Hundhausen M, Ley L (1995) Appl Phys Lett 67:3144
443. Gordon AE, Fayfield RT, Litfin DD, Higman TK (1995) J Vac Sci Technol B 13:2805
444. Tsau L, Wang D, Wang KL (1994) Appl Phys Lett 64:2133
445. Avouris P, Hertel T, Martel R (1997) Appl Phys Lett 71:285
446. Kim Y-T, Bard AJ (1992) Lanmuir 8:1096
447. Stockman L, Neuttiens G, van Haesendonck C, Bruynseraede Y (1993) Appl Phys Lett 62:2935
448. Schoer JK, Ross CB, Crooks RM, Corbitt TS, Hampden-Smith MJ (1994) Langmuir 10:615
449. Marrian CRK, Perkins FK, Brandow SL, Koloski TS, Dobisz EA, Calvert JM (1994) Appl Phys Lett 64:390
450. Xu LS, Alleee DR (1995) J Vac Sci Technol B 13:2837
451. Yaniv DR, MaCormick LD (1992) Nanotechnology 3:44
452. Sasano K, Nakamura K, Kaneto K (1993) Jpn J Appl Phys 2 32:L863
453. Yang R, Evans DF, Hendrickson WA (1995) Langmuir 11:211
454. Borgwarth K, Ricken C, Ebling DG, Heinze J (1995) Bunsen Ges Phys Chem 99:1421
455. Nyffenegger R, Ammann E, Siegenthaler H, Kötz R (1995) Electrochim Acta 40:1411
456. Tang SL, McGhie AJ, Lewittes ME (1992) Appl Phys Lett 60:1821
457. Wendel M, Kuhn S, Lorenz H, Kotthaus JP, Holland M (1994) Appl Phys Lett 65:1775
458. Göbel H, van Blanckenhagen P (1995) J Vac Sci Technol B 13:1247
459. Goto K, Hane K (1996) Rev Sci Instrum 67:397
460. Jin X, Unertl WN (1992) Appl Phys Lett 61:657
461. Yamamoto S-I, Yamada H, Tokumoto H (1995) Jpn J Appl Phys 34:3396
462. Gorwadkar S, Vinogradov GK, Senda K, Morita S (1995) J Appl Phys 78:2242
463. Khurshudov A, Kato K (1995) J Vac Sci Technol B 13:1938
464. Jung TA, Moser A, Hug HJ, Brodbeck D, Hofer R, Hidber HR, Schwarz DU (1992) Ultramicroscopy 42–44:1446
465. Wiesendanger R (1995) Jpn J Appl Phys 34:3388
466. Jaschke M, Butt H-J (1995) Langmuir 11:1061
467. Bouju X, Joachim C, Girard C (1994) Phys Rev B 50:7893
468. Junno T, Deppert K, Montelius L, Samuelson L (1995) Appl Phys Lett 66:3627
469. Müller WT, Klein DL, Lee T, Clarke J, McEuen PL, Schultz PG (1995) Science 268:272
470. Bhushan B, Israelachvili JN, Landman U (1995) Nature 374:607
471. Feynman RP (1960) Engineering and Science Feb:22
472. Weissbuch I, Baxter PNW, Cohen S, Shubert US, Lehn J-M, Leiserowitz L (1998) J Am Chem Soc 120:4850
473. Whitesides GM, Simanek EE, Mathias JP, Seto CT, Chin DN, Mammen M, Gordon DM (1995) Acc Chem Res 28:37
474. McGrath KP, Fournier MJ, Mason TL, Tirell DA (1992) J Am Chem Soc 114:727
475. Hansma HG (1996) J Vac Sci Technol B 14:1390
476. Bustamante C, Rivetti C, Keller DJ (1997) Curr Opin Struct Biol 7:709
477. Fritzsche W, Takac L, Henderson E (1997) Crit Rev Eukaryotic Gene Expression 7:231
478. Engel A, Schoenenberger CA, Müller DJ (1997) Curr Opin Struct Biol 7:279
479. Micas. Reviews in Mineralogy (1984) Bailey SW (ed) Chelsea Book Crafters, Inc.
480. Pashley RM (1981) J Colloid Interface Sci 80:153
481. Bustamante C, Rivetti C (1996) Annu Rev Biophys Struct 25:395
482. Bustamante C, Keller D, Yang G (1993) Curr Opin Struct Biol 3:363
483. Lyubchenko YL, Oden PI, Lampner D, Lindsay SM, Dunker KA (1993) Nucleic Acids Research 21:1117

484. Schaper A, Starink JP, Jovin TM (1994) FEBS Lett 355:91
485. Hansma HG, Bezanilla M, Zenhausern F, Adrian M, Sinsheimer RL (1993) Nucleic Acids Research 21:505
486. Zenhausern F, Adrian M, ten Heggeler-Bordier B, Eng LM, Descouts P (1992) Scanning 14:212
487. Hansma HG, Vesenka J, Siegerist C, Kelderman G, Morrett H, Sinsheimer RL, Elings V, Bustamante C, Hansma PK (1992) Science 256:1180
488. Yang G, Vesenka JP, Bustamante CJ (1996) Scanning 18:344
489. Hansma HG, Sinsheimer RL, Groppe J, Bruice TC, Elings V, Gurley G, Bezanilla M, Mastrangelo IA, Hough PVC, Hansma PK (1993) Scanning 15:296
490. Leuba S, Yang G, Robert C, Samorí B, van Holde K, Zlatanova J, Bustamante C (1994) Proc Natl Acad Sci USA 91:11621
491. Han W, Lindsay SM (1997) Nature 386:563
492. Hallett P, Tskhovrebova L, Trinick J, Offer G, Miles MJ (1996) J Vac Sci Technol B 14:1444
493. Rivetti C, Guthold M, Bustamante C (1996) J Mol Biol 264:919
494. Kratky O, Porod G (1949) Recueil 68:1106
495. Schellman JA (1974) Biopolymers 13:217
496. Landau LD, Lifshitz EM (1980) Statistical Physics, Part 1. 3rd Ed, Pergamon Press, Oxford, NY
497. Grosberg AY, Khokhlov AR (1994) Statistical Physics of Macromolecules. API Press, Woodbury, NY
498. Bustamante C, Marko JF, Siggia ED, Smith S (1994) Science 265:1599
499. Vasquez C, Kleinschmidt AK (1968) J Mol Biol 34:137
500. Butt H-J, Downing KH, Hansma PK (1990) Biophys J 58:1473
501. Hoh JH, Sosinsky GE, Revel J-P, Hansma PK (1993) Biophys J 65:149
502. Singh S, Turina P, Bustamante C, Keller DJ, Capaldi R (1996) FEBS Lett 397:30
503. Takeyasu K, Omote H, Nettikadan S, Tokumasu T, Iwamoto-Kihara A, Futai M (1996) FEBS Lett 392:110
504. Mou J, Czajkowsky DM, Sheng SJ, Ho R, Shao Z (1996) FEBS Lett 381:161
505. Mou J, Sheng S, Ho R, Shao Z (1996) Biophys J 71:2213
506. Braunstein D (1995) J Vac Sci Technol A 13:1733
507. Zhu J, Hartman J, Case R, Rice S, Vale R (1998) Appl Note of Molecular Imaging Inc. http://www.molec.com
508. Haggerty L, Watson BA, Barteu MA, Lenhoff AM (1991) J Vac Sci Technol B 9:1219
509. Yoshimura H (1997) Adv Biophys 34:93
510. Müller DJ, Schabert FA, Müldt G, Engel A (1996) Biophys J 68:1681
511. Müller JD, Baumeister W, Engel A (1996) J Bacteriol 178:3025
512. Podgornik R, Strey HH, Gawrisch K, Rau DC, Rupprecht A, Parsegian VA (1996) Proc Natl Acad Sci USA 93:4261
513. Fang Y, Yang J (1997) J Phys Chem B 101:441
514. Vater W, Bohm KJ, Unger E (1997) Cell Mobility an the Cytoskeleton 36:76
515. Anselmetti D, Dreier M, Lüthi R, Richmond T, Mayer E, Frommer J, Guntherodt H-J (1994) J Vac Sci Technol B 12:1500
516. Maeda H (1997) Langmuir 13:4150
517. Wadu-Mesthrige K, Pati B, McClain M, Liu G-Y (1996) Langmuir 12:3511
518. Kralchevsky K, Nagayama (1994) Langmuir 10:23
519. Hansma HG, Laney DE, Revenko I, Kim K, Cleveland JP (1996) in Biological Structure and Dynamics. Sarma RH, Sarma MH (ed) Adenine Press, Albany NY
520. Lyubchenko YL, Shlyakhtenko LS (1997) Proc Natl Acad Sci USA 94:496
521. Rippe K, Mucke N, Langowski J (1997) Nucleic Acids Res 25:1736
522. Thomson NH, Fritz M, Radmacher M, Cleveland JP, Schmidt CF, Hansma PK (1996) Biophys J 70:2421
523. Erie D, Bustamante C (1995) Science 269:989

524. Wyman C, Rombel I, North AK, Bustamante C, Kustu S (1997) Science 275:1658
525. Kasas S, Thomson NH, Smith BL, Hansma HG, Zhu X, Guthold M, Bustmante C, Kool ET, Kashlev M, Hansma PK (1997) Biochemistry 36:461
526. Zhu YJ, Han W, Dlakic M, Lindsay SM, Harrington RE (1998) Appl Note of Molecular Imaging Inc. http://www.molec.com
527. Roovers J J (1991) Non.-Cryst. Solids 793:131
528. Gauthier M, Möller M (1991) Macromolecules 24:4548
529. Hempenius MA, Michelberger W, Möller M (1997) Macromolecules 30:5602
530. Wu J, Lieser G, Wegner G (1996) Adv Materials 8:151
531. Wintermantel M, Fischer K, Gerle M, Ries R, Schmidt M, Kajiwara K, Urakawa H, Wataoka I (1995) Angew Chem Int Ed Engl 34:1472
532. Beers KL, Gaynor SG, Matyjaszewski K, Sheiko SS, Möler M (1998) Macromolecules 31:9413
533. Karakaya B, Claussen W, Gessier K, Säanger W, Schlüter A-D (1997) J Am Chem Soc 119:3296
534. Yin R, Zhu Y, Tomalia DA (1998) J Am Chem Soc 120:2678
535. Zimmerman SC (1997) Curr Opin Coll Interf Sci 2:89
536. Zhao M, Tokuhisa H, Crooks RM (1997) Angew Chem Int Ed Engl 36:2596
537. Cooper AJ, Londono JD, Wignall G, McClain JB, Samulski ET, Lin JS, Dobrynin A, Rubinstein M, Burke ALC, Frechet JMJ (1997) Nature 389:368
538. Jansen JFGA, de Brabender-van der Berg EMM, Meijer EW (1995) J Am Chem Soc 117:4417
539. Zhao M, Crooks RM (1999) Adv Mater 11:217
540. Deckman HW, Dunsmuir JH (1988) J Vac Sci Technol B 6:333
541. Burmeister F, Schaefle C, Mattes T, Boehmisch M, Boneberg J, Leiderer P (1997) Langmuir 13:2983
542. Wang PW, Liu YJ, Devadoss C, Bharathi P, Moore JS (1996) Adv Mater 8:237
543. Dandliker J, Diederich F, Gisselbrecht J-P, Louati A, Gross M (1995) Angew Chem Int Ed Engl 34:2725.
544. Albrecht M, van Koten G (1999) Adv Mater 11:171
545. Vasilenko NG, Rebrov EA, Muzafarov AM, Eßwein B, Striegel B, Möller M (1998) Macromol. Chem. Phys 199:889
546. Leduc MR, Hawker CJ, Dao J, Frechet JMJ (1996) J Am Chem Soc 118:11111
547. Burmeister F, Schaefle C, Keilhofer B, Bechinger C, Boneberg J, Leiderer P (1998) Adv Mater 10:495
548. Fendler JH (1996) Curr Opin Coll Interf Sci 1:202
549. Chevalier Y, Pichot C, Graillat C, Joanicot M, Wong K, Maquet J, Lindner, Cabane B (1992) Colloid Polym Sci 270:806
550. Dimitrov AS, Dushkin CD, Yoshimura H, Nagayama K (1994) Langmuir 10:432
551. Micheletto R, Fukuda H, Ohtsu M (1995) Langmuir 11:3333
552. Winnik MA (1997) Curr Opin Coll Interface Sci 2:192
553. Linemann RF, Malner TE, Brandsch R, Bar G, Ritter W, Mülhaupt R (1999) Langmuir in press
554. Antonietti M, Basten R, Lohmann S (1995) Macromol Chem Phys 196:441
555. Johnson CA, Lenhoff AM (1996) J Coll Interface Sci 179:587
556. Stamou D, Liley M, Vogel H, Duschl C (1998) Polymer Prepr Am Chem Soc Polym Div 39(2):1161
557. Semmler M, Mann EK, Ricka J, Borkovec M (1998) Langmuir 14:5127
558. Adamczyk Z, Warszynski P (1996) Adv Coll Interf sci 63:41
559. Mourey TH, Turner SR, Rubinstein M, Fréchet JMJ, Hawker CJ, Wooley KL (1992) Macromolecules 25:2401
560. Gauthier M, Möller M, Burchard W (1994) Macromol Symp 77:43
561. Hawker CJ, Farrington PJ, Mackay ME, Wooley KL, Fréchet JMJ (1995) J Am Chem Soc 117:4409

562. Hempenius MA, Zoetelief WF, Gauthier M, Möller M (1998) Macromolecules 31:2299
563. Jackson CL, Chanzy HD, Booy FP, Drake BJ, Tomalia DA, Bauer BJ, Amis EJ (1998) Macromolecules 31:6259
564. Coen MC, Lorenz K, Kressler J, Frey H, Mülhaupt R (1996) Macromolecules 29:8069
565. Wells M, Crooks RM (1996) J Am Chem Soc 118:3988
566. Saville PM, White JW, Hawker CJ, Wooley KL, Fréchet JMJ (1993) J Phys Chem 97:293 ibid. (1995) 99:8283
567. De Leuze-Jallouli AM, Swanson D, Dvornic PR, Perz SV, Owen MJ (1997) PMSE Proceedings 77:93
568. Sheiko SS, Buzin AI, Muzafarov AM, Rebrov EA, Getmanova EV (1998) Langmuir 14:7468
569. Tsukruk VV, Rinderspacher F, Bliznyuk VN (1997) Langmuir 13:2171
570. Takada K, Diaz DJ, Abruna HD, Cuadrado I, Casado C, Alonso B, Moran M, Losada J (1997) J Am Chem Soc 119:10763
571. Watanabo S, Regen SL (1994) J Am Chem Soc 116:8855
572. Hierlemann A, Campbell JK, Baker LA, Crooks RM, Ricco AJ (1998) J Am Chem Soc 120:5323
573. Huck WTS, van Veggel FCJM, Sheiko SS, Möller M, Reinhoudt DN (1998) J Phys Org Chem 11:540
574. Mansfield ML (1996) Polymer 37:3835
575. Cao Y, Smith P (1993) Polymer 34:3139
576. Ruokolainen J, Torkkeli M, Serimaa R, Vahvaselkä S, Saariaho M, ten Brinke G (1996) Macromolecules 29:6621
577. Birshtein TM, Borisov OV, Zhulina EB, Khokhlov AR, Yurasova TA (1987) Polymer Sci. USSR 29:1293
578. Fredrickson GH (1993) Macromolecules 26:2825
579. Saariaho M, Ikkala O, Szleifer I, Erukhimovich I, ten Brinke G (1997) J Chem Phys 107:3267
580. Wintermantel M, Schmidt M, Tsukahara Y, Kajiwara K, Kohjiya S (1994) Macromol Rapid Comm 15:279
581. Gerle M, Fischer K, Schmidt M, Roos S, Müller AHE, Sheiko SS, Prokhorova SA, Möller M (1999) Macromolecules 32:2629
582. Sheiko SS, Gerle M, Fischer K, Schmidt M, Möller M (1997) Langmuir 20:5368
583. Williams DRM (1993) J.Phys.II (France) 3:1313
584. Rouault Y, Borisov OV (1996) Macromolecules 29:2605
585. Soga KG, Guo H, Zuckermann MJ (1995) Europhys Lett 29:531
586. Zhulina E, Singh C, Balazs AC (1996) Macromolecules 29:6338
587. Siqueira DF, Kohler K, Stamm M (1995) Langmuir 11:3092
588. Stamouli A, Pelletier E, Koutsos V, van der Vegte E, Hadziionaou G (1996) Langmuir 3221
589. Sheiko SS, Borisov OV, Prokhorova SA, Schmidt U, Gerle M, Schmidt M, Möller M submitted to Phys Rev Lett
590. Dobrynin AV, Rubinstein M, Obukhov SP (1996) Macromolecules 29:2974
591. Sheiko SS, Prokhorova SA, Schmidt U, Dziezok P, Schmidt M, Möller M (1999) in ACS Symposium Series 741, Tsukruk VV, Wahl K (ed) ACS Washington (to be published)
592. Balagurusamy VSK, Ungar G, Percec V, Johansson G (1997) J Am Chem Soc 119:1539
593. Percec V, Ahn C-H, Ungar G, Yeardley DSP, Möller M, Sheiko SS (1998) Nature 391:161
594. Prokhorova SA, Sheiko SS, Möller M, Ahn C-H, Percec V (1998) Macromol Rapid Comm 19:359
595. Stocker W, Schürmann BL, Rabe JP, Förster S, Lindner P, Neubert I, Schlüter A-D (1998) Adv Mater 10:793
596. Prokhorova SA, Sheiko SS, Möller M, Ahn C-H, Percec V (1999) Macromolecules 32:2653

597. Kwon YK, Chvalun S, Scheider A-I, Blackwell S, Percec V, Heck JA (1994) Macromolecules 27:6129
598. Marko, J.F.; Siggia, E.D. Science 1994, 265:506
599. Marko, J.F.; Siggia, E.D. Physical Review E 1995, 52, 2912
600. McLeish, T. Science 1997, 278, 1577
601. Strick TR, Allemand J-F, Bensimon D, Croquette V (1998) Biophys J 74:2016
602. Rabe JP Buchholz S (1991) Phys Rev Lett 66:2096
603. Wavkuschewskij A, Cantow H-J, Magonov SN, Möller M, Liany W, Whanbo M-H (1995) Adv Mater 5:821
604. Bauerle P, Fischer T, Bidlingmeier B, Stabel A, Rabe JP (1995) Angew Chem Int. Ed Engl 34:303
605. Merkel R, Nassoy P, Leung A, Ritchie K, Evans E (1999) Nature 397:50
606. Cluzel P, Lebrun A, Heller C, Lavery R, Viovy J-L, Chatenay D, Caron F (1996) Science 271:792
607. Smith SB, Finzi L, Bustamante C (1992) Science 258:1122
608. Perkins TT, Quake SR, Smith DE, Chu S (1994) Science 264:822
609. Wang MD, Yin H, Landick R, Gelles J, Block SM (1997) Biophys J 72:1335
610. Baumann CG, Smith SB, Bloomfield VA, Bustamante C (1997) Proc Nat Acad Sci 94:6185
611. Leckband DE, Israelachvili JN, Scmitt F-J, Knoll W (1992) Science 255:1419
612. Hoh JN, Cleveland JP, Prater CB, Revel J-P, Hansma PK (1993) J Am Chem Soc 114:4917
613. Radmacher M, Fritz M, Hansma HG, Hansma PK (1994) Science 265:1577
614. Perkins TT, Smith DE, Larson RG, Chu S (1995) Science 268:83
615. Schurr JM, Smith SB (1990) 29:1161
616. Vologodskii A (1994) Macromolecules 27:5623
617. Marko JF, Siggia ED (1995) Macromolecules 28:8759
618. Moroz JD, Nelson P (1998) Macromolecules 31:6333
619. Lee GU, Kidwell DA, Colton RJ (1994) Langmuir 10:354
620. Lee GU, Chrisey LA, Colton RJ (1994) Science 266:771
621. Hendrikson WA, Pahler A, Smith JL, Satow Y, Merritt EA, Philazackerley RP (1989) Proc Natl Acad Sci USA 86:2190
622. Rief M, Oesterhelt F, Heymann B, Gaub HE (1997) Science 275:1295
623. Li H, Rief M, Oesterhelt F, Gaub HE (1998) Adv Mater 3:316
624. Labeit S (1990) Nature 345:273
625. Pan KM, Damodaran S, Greaser ML (1994) Biochemistry 33:8255
626. Rief M, Gautel M, Oesterhelt F, Fernandez Jm, Gaub HE (1997) Science 276:1109
627. Kellermayer MSZ, Smith SB, Granzier HL, Bustamante C (1997) Science 276:1112
628. Nave R, Fürst DO, Weber K (1989) J Cell Biol 109:2177
629. Evans E, Ritchie K (1997) Biophys J 72:1541
630. Thundat T, Allison DP, Warmack RJ (1994) Nucleic Acids Res 22:4224
631. Bensimon D, Simon AJ, Croquette V, Bensimon A (1995) Phys Rev Lett 74:4754
632. Uchida E, Ikada Y (1997) Macromolecules 30:5464
633. Koutsos V, van der Vegte EW, Pelletier E, Stamouli A, Hadziioannou G (1997) Macromolecules 30:4719
634. Stamouli A, Pelletier E, Koutsos V, van der Vegte E, Hadziioannou G (1996) Langmuir 12:3221
635. Ortiz C, Hadziioannou G (1999) Macromolecules 32:780
636. Tang H, Cuberes MT, Joachim C, Gimzewski JK (1997) Surface Sci 386:115
637. Gimzewski JK, Jung TA, Cuberes MT, Schlittler RR (1997) Surface Sci 386:101
638. Heller EJ, Cromie MF, Lutz CP, Eigler DM (1994) Nature 369:464
639. Crommie MF, Lutz CP, Eigler DM, Heller EJ (1996) Surf Sci 361–362:864
640. Yazdani A, Eigler DM, Lang ND (1996) Science 272:1

Received: April 1999

Author Index Volumes 101–151

Author Index Volumes 1–100 see Volume 100

Subject Index

MIX
Papier aus verantwortungsvollen Quellen
Paper from responsible sources
FSC® C105338

If you have any concerns about our products,
you can contact us on
ProductSafety@springernature.com

In case Publisher is established outside the EU,
the EU authorized representative is:
Springer Nature Customer Service Center GmbH
Europaplatz 3, 69115 Heidelberg, Germany

Printed by Libri Plureos GmbH
in Hamburg, Germany